LAND QUESTION IN TRIPURA

LAND QUESTION IN TRIPURA

DIPANNITA CHAKRABORTY

AKANSHA PUBLISHING HOUSE
NEW DELHI-110059 (INDIA)

AKANSHA PUBLISHING HOUSE
R 37-B, Vani Vihar, Uttam Nagar
New Delhi-110059
Email: ektabooks@yahoo.com
Ph. : 56065579 / *9811582579*

Land Question in Tripura

First Edition 2004
ISBN 81–87606–57-6

PRINTED IN INDIA

Published by M.P. Misra for Akansha Publishing House, New Delhi and Printed at Tarun Offset Printers, Delhi-53

Preface

Tripura, a tiny state in the North-Eastern Region of India has an area of 10,491 square kilometres and a large population base comprising of 31,91,168 persons (according to the 2001 Census). The economy of the state is heavily dominated by the primary sector and hence population pressure on land is inevitable which manifests in decline in the size of holdings. There existed a dual cultivation method since the previous century, shifting in the hills and settled in the plains. A harmonious blending of these two methods as well as those who practised them was a noticeable feature of the state. But this honeymoon was shortlived. The British overlords who till 1850s did not bother about this hilly state started showing much interest in its strategic location during the late 19th century. They craved for higher revenue returns and so introduced certain administrative reforms. Agriculture, the mainstay of the population of the state was moulded accordingly and new settlements encouraged which expanded plough cultivation in the vast fertile plain tracts available, though shifting remained the cultivators had to retreat to remoter parts of the state to pursue the same.

However, the partition of the sub-continent and the following upheavals led to an unprecedented rise in the state population resulting in an imbalance in the agrarian structure of the state, which gave vent to social, economic and political problems arising out of land-related issue. Land which was abundant a few years back gradually shrunk rendering a large number of people landless, productivity declined as population increased unchecked and the state economy was thrown out of gear.

The present study investigates into the causes which led to this low availability of cultivable land both in the hills and the plains, in an historical perspective. It highlights the various manifestations of the land systems evolved by the princely regime and thereafter by the subsequent state government and analyses its impact on land. The time horizon spreads over 19th and 20th centuries, but the early 19th century witnessed the predominance of the primitive cultivation in the Hill Tipperah region which involved neither land tenure system nor any individual ownership right on land. The crisis of land availability cropped up when population began to rise i.e. towards the end of the century and plough cultivation spread over extensive areas and land began to be individually owned.

I owe a great deal to several persons and institutions whose exhaustive list is impossible to draw up here. I am greatly indebted to Dr. H.K. Sur, Reader, Tripura University, Agartala without whose able guidance this pursuit could not have been completed.

I express my deep sense of gratitude to Dr. Mahadev Chakravarti, Professor, Department of History, Tripura University and Dr. D.K. Chaudhuri, Registrar, Tripura University for their valuable advice and vast stock of information so generously placed at my disposal.

I am much thankful to Dr. Amalendu De, Guru Nanak Professor of Indian History (Retd.), Jadavpur University for his comments and encouragement given to me during the public seminar of the present work.

I am thankful to the Librarians of the National Library, Calcutta, Asiatic Society, Calcutta, Tripura University, Agartala. The Tribal Research Centre, Agartala and the Tripura Legislative Assembly, Agartala for providing necessary library facilities. I also acknowledge my gratitude to the staff of the State Archives of West Bengal, Calcutta and the Secretariat Archives, Agartala who helped me in collecting all the available reports of the Political Agencies as well as some rare documents, notifications, official promulgations etc. Thanks are due to several officers, departmental heads and staff of the Directorate of Land Records and Settlement, Land Use Board, Forestry, Rubber Plantation Corporation, Refugee Rehabilitation, Statistics, Tripura Tribal Areas Autonomous District Council, Land Acquisition and Land Revenue Departments for their unstinted cooperation extended to me.

I am greatly indebted to Sri K.K. Chakraborty, Assistant Teacher, Barpathari High School with whose help the field studies in the Belonia sub-division could be conducted smoothly within a short span. My special thanks are to Dr. M.L. Das, Asstt. Professor, Women's College, Agartala, who, despite his busy schedule helped me to complete the thesis.

In the end, I express my gratitude to all those who have directly or indirectly helped me in the completion of the pursuit.

DIPANNITA CHAKRABORTY

Contents

List of Tables

1

Introduction

Prelude

Land, in a fundamental sense is the most valuable endowment of nature, as it meets the two basic requirements of mankind i.e. food and shelter. Besides providing us with forests which maintains the ecological balance; land caters to our non-agricultural needs too. Importance of land in a historical perspective may be traced to the time when early man first began to utilise land for his livelihood. He needed food to eat so learnt to grow crops and giving up his nomadic nature, settled on a plot of land which was till then a general property of mankind. Land now became a transient property for him who used it and thus acquired a sort of ownership. Next, the early man decided to pursue agriculture, an activity by which he sought to make use of available land and tried to accelarate and improve upon the natural growth so that the process helped him to yield more crops. Land, during this time had a two dimensional use since only the area (i.e. length and breadth of the surface) was actually needed for pastures, agriculture and settlement. A third dimension was added since the Industrial Revolution when the minerals beneath the surface of the earth acquired much importance. When mankind increased in number a conception of domination—a permanent ownership pattern on land ensued wherein the individual could own it for a long period uninterruptedly hence rights over this natural resource was acquired. Agriculture has remained with human civilization all along, and population has affected the land utilisation pattern at the same time, the relation between these two factors makes the subject matter for land question.

Land, according to its uses, can be divided into several categories like the flora and fauna, agriculture, forests, mines and settlements (industries, roads, towns etc.). Of these, cultivable land is quite limited in supply as not all available land is cultivable. In fact, only a third of the land surface is actually cultivable. The cultivability of land depends upon the soil temperature, water availability and landforms which are the physical frontiers of land. The natural vegetation of land too determines its cultivability since it restores fertility as well as arrests erosion. With the progress of time, man by his will and wisdom has been able to increase production of this limited resource by adopting several means viz., irrigation, manure, high quality seeds, technology etc. as well as reclaimed fallow land. Several studies on farm management have shown that there is an inverse relationship between the size of the farm and its productivity, hence land reform enactments have come into effect. Here too rural land is regarded as important because of its resource base (timber and food production) which is strategic for agricultural self sufficiency. With industrialisation, urban use of land gained magnitude and such planning pattern overlapped at times. Land is thus, an essential factor in the human society subject to its uses and with civilization the dimensions of land use keeps on increasing. Recognition of the value of land in its heritage context and in ecological terms has given rise to an expanding interest in the conservation of land which has become a point of great concern not only in deep rural areas but also on the urban fringes. Land, it should be remembered is to be protected, managed and developed too so that it is safeguarded as the basis for human life and environment. Hence, the area available to a region, coupled with the pattern of utilization is of much importance as the need of the hour is optimum use of land for maximum benefit.

In India, agriculture stands to be the mainstay for millions of its inhabitants as industrialisation is still a far cry. Even in the last lap of 20th century any major shift of the population from agriculture to industry seems bleak which could have released the pressure on land to a considerable extent. The north-eastern region of the country is nature's veritable arcadia but its hilly terrain and certain complexities (relating to socio-cultural life, traditional knowledge, technology, infrastructure etc.) have made it lag behind the other states of the country in this context, making the situation rather grim. It adjoins three countries—India, China and Myanmar in the east and flanks two other countries, Bhutan and Bangladesh on the west. It is a geographical conglomerate of seven states namely, Arunachal Pradesh, Assam, Manipur, Mizoram, Meghalaya, Nagaland and Tripura inhabited by Tibeto-Mongoloids. This sector has a combined potential of becoming the richest segments of the country yet it remains under developed. Independence saw the North-east as a war ravaged, economically unstable, politically weak, but ethnically potent area and ideal for development with a huge natural resource base but apathetic treatment by the centre and disinterest shown by the state government have been culpable for non-development. North East India accounts for 7.76 per cent of the geographical area of the country where the hills constitute about 70 per cent of the total land area.[1] Forests, which have a definite claim on the available land are spread over 46.9 per cent of the area.[2] Agriculture is the predominant occupation of the people and shifting cultivation (locally known as Jhum) is widely practised. About 5.7 per cent of geographical area remains under shifting cultivation at one time or the other involving a total of 4,43,336 families.[3] The population of the region is 31,547,314 accordingly to 1991 census which has grown by 2.38 per cent during the last decade where as that of the country was–0.87 per cent portraying the fact that the growth is much above the national average.[4] The pressure on land, thus, is very high and is rising further as a consequence of growing population in this region particularly in Assam and Tripura.[5] Such a trend obviously has an adverse impact on the north-eastern economy where hills form a major share in the topography of the area. Several man-made activities viz. road-making, developmental projects, habitation, institutions, security purposes etc. have grown along time taking up large chunks of land area leaving a net cultivated area of 14,75,000 hectares (All India Bulletin on Agricultural Census 1985-

86. Basic Statistics 1991) only which is quite low in proportion to the population of the region, thus explaining the high density of population.[6] These factors, taken together reveals the paucity of land available in the region for cultivation purposes and thus accounts for meagre land productivity which hardly suffices for subsistence even, putting forth the moot point of land question. The situation in Tripura, one of the seven states of the north-east poses certain distinct complexities regarding the land question of the State.

The state of Tripura, formerly a feudal state, is essentially a frontier region. Except for an irregular strip to the West and South and smaller pockets of flat lands entering the hills to the North West, it consists of a mountain range between Brahmaputra and Chindwin Irrawady river system. A tiny and hilly state Tripura physiographically represents the western fringe of the typical 'ridge and valley' province commonly known as Indo-Burman Ranges. Hence geo-strategically it is quite a vital cardinal area of the nation. Like the other north-eastern states, its topography too is undulating and characterised by numerous hillocks, plains, valleys, lungas, rivers, rivulets etc. other than the vast forest cover. The flat land constitutes 40 per cent of the total area and accommodates two-thirds population. The state experienced spurts of population growth at one time or another, but the interesting fact is that Tripura's growth rate has always maintained a higher rate, higher than that of the country even during the first half of the present century. While India's percentage increase of population was 51.50 that of Tripura's was an astounding 273.41[7]. Being predominantly an agricultural state, this unusual growth of population obviously raised eyebrows regarding accommodation of the people both physically and economically. It should also be borne in mind that Tripura's population, like the other states of the North-east, is not of the same ethnicity. The kings invited the plainlanders initially but soon this invitation took the shape of mass influx following political turmoils which put the agrarian economy of the state in a dismal condition. The autochthons felt discriminated against so reserves were created while at the same time the displaced people needed shelter so rehabilitation projects were established. Forest land had to be conserved for maintaining ecological balance, simultaneously land was needed for accommodating the people. Autonomous Council was set up which reserved quite a huge chunk of the land which had a tribal majority and this time the plainlanders felt discriminated against. Widespread shifting cultivation rendered a vast region uncultivable due to loss of soil fertility, erosion and to top it all, land alienation took place in various ways. This has made the state's land question unique in comparison to the rest of the region.

A Glimpse into the Annals of Tripura

Tripura was earlier known as 'Independent Tipperah' and later as Hill Tipperah' till 1920, a nomenclature earned during the British period owing to its physiognomy. Its name was changed to Tripura at the request of the rulers. It possessed a total area of 3870 sq. miles[8] and the hills covered most of this area. The hills commenced from the western side and gradually rose towards the eastern side, the valleys in between were broken up by rough and irregular ridges of low hills. Dense forests covered the hills while the valleys were swampy. Geographically, the state was attached to Samatata or East Bengal and did not have any independent status, since the ancient period.

The history of the state during the previous centuries was intertwined with East Pakistan, a region roughly corresponding to the present district of Sylhet, Noakhali & Chittagong[9]. The king of Tipperah possessed the plains which included Sarail, Bardakhat, Gangamandal and Pattikera districts along with Chakla Roshanabad[10]. However, these districts except Chakla Roshanabad were lost to them as an aftermath of family feuds. Turning back the pages of history, it is known that Vainya Gupta, a descendent of the Guptas is said to have made a land grant in Tripura[11] (6th century A.D.). There are also epigraphic evidences showing the rule of Khadgas[12] (late 7th century A.D.) followed by the Ratas and Devas, till 10th century A.D.[13] Then ruled the Chandras[14] while the Palas controlled the regions between Lalmai-Mainamati hills till the outbreak of Kaivarta revolt. The Varmans, who ruled Dhaka-Faridpur extended their sway over the land. Thus we find that most of the western and southern parts of the state were politically attached to East-Bengal under various dynasties in the ancient times. The Tipra tribe were confined in the hilly regions. At and around the same time, one Jujharu Fa conquered Rangamati (Udaipur) and Bishalgarh, defeating Samachardeva a Deva ruler of Samatata. The conquerer belonged to the Tipra tribe and gradually spread his authority in the adjoining areas, where he grew powerful and extended political influence; he even gave up the title of 'Fa' (15th century) being influenced by Hindu society and culture[15], thus started a close relationship between the state of Tripura and the neighbouring Samatata (East Pakistan). The next noted king of the land, Ratna—Fa, is said to have requested the Sultan of Gaud (Barbak Shah) to send some Bengali families to his state, which again confirms a change in the royal social outlook. About 4,000 Bengali families[16] are said to have settled in Tripura, of whom some even became ministers by dint of their capability.

However, with the coming of the Mughals, the kingdom of Tripura received a set back as Udaipur was occupied (1584), garrisoned (1618) and later converted into a Sarkar[17]. Interestingly, the low land was captured and not the entire kingdom, as it could fetch regular rents, and the hills were left to the kings. The Muslim Zamindars, who were entrusted with the plainsland often interfered in the state administration in making and unmaking kings. In the Revenue Records of Bengal Suba (1658) Sarkar Udaipur was recorded as a revenue paying centre.[17] Meanwhile, a revenue clerk of Chakla Roshanabad region became the virtual ruler of the area by defeating the king of Tripura in 1748 A.D.[18] The revenue of Chakla Roshanabad increased manifold as he tried to introduce administrative reforms and regulations. He even granted rent-free land to the Hindu and Muslim inhabitants of Roshanabad. However opposition from the royalists brought about his downfall in 1760. The British followed suit, after taking over the administrative reins of Bengal (Burdwan, Midnapur and Chittagong) in accordance to a secret treaty with Mir Jafar, the Nawab of Bengal. They cast their covetous eyes on Tripura, Verelst sent a troop (1761) against king Krishna Manikya in answer to the Nawab's call for help and the king gave in without a fight. The British flag flew over the plains of Tripura and the revenue was directed towards the British exchequer, the king was required to bear the troop's maintenance charge too.[19] Curiously, Tripura never had any treaty relations with the British, though they had acquired the privilege of "investing" the royal heir to the throne with the insignia of kingship since 1809.[20] A controversy rose regarding the right of the king to levy transit duties on hill produces, wherein the Deputy Governor of Bengal clarified the fact that the king of Tripura has full right within his hill territory to levy any duty he pleases, thus stressing on the independent status of the king of Tripura even during the hey day of the Raj.[21] Independent status of the kings of the state was proved as the kings got a free hand as to taxation, absolute powers over subjects etc. but such independence was practically dependent in submission to the British paramount power.

Land Systems During the 19th and 20th Centuries: An Overview

The ruling elite of the state, though considered "Independent" in their own hilly territory were often embroiled in border disputes since long. Previous to the Muslim conquest of the Northern portion (Sylhet) of the state the Tipperah kingdom is said to have extended over the valleys which were covered mostly with dense forest and impassable swamps. Cultivation which was quite extensive earlier was stopped during 19th century perhaps due to pestilence, wild beasts or petty warfare and the cultivated land again turned into swamps.[22] Cultivations were spread with Muslim conquest and the kings asserted their rights on the land giving rise to land-disputes. Arbitrarily large chunks of 'settled' land was taken away by the British on the flimsy plea of unauthentic documents. The Kings had already granted these lands to several tenants as talooks, but inability to prove that the lands were not included in the arbitration the claims of the king on such land were dismissed. Maharaja Birendra Kishore Manikya naturally felt dejected as is seen from his letter to Sir William Marris, Governor of Assam where he laments after the boundary demarcations—"The land has not been wrested from the Rulers of Tripura by right of conquest of annexation for any misrule or misconduct on their part, but gradually absorbed in British India for reasons unknown to the Durbar."[23]

Conspicuously enough the hills were left untouched by the British overlords, perhaps they did not find them worthy from commercial point of view yet, being covered with forests and marshes. The kings had to fall back upon these hitherto neglected hills of the state with mounting pressure of revenue exactions. Though land was abundant, there were hardly any takers as no proper land tenure was evolved and the raiyats (tenants) could not get proper occupancy rights, or any guarantee even which they could secure under British Government elsewhere.[24] The King now began leasing out the hills and available plains to the indigenous chiefs (Thakurs) in lieu of low or no rent at all. Land in the British district which belonged to the kings of Tripura were under exclusive rights of the British who settled it themselves. The chiefs in the Tripura state later distributed the land in the hills to the autochthons or other tribes who pursued the rudimentary method of shifting cultivation, while the land on the plains were sub-let to the neighbouring Bengali or Manipuri cultivators at high rentals.

In a land abundant thinly populated state like Tripura, the Manikyas introduced a revenue oriented tenurial system without taking into consideration either its development or consequences.[25] In due course land revenue formed the greatest source of wealth for the rulers. The rate of revenue varied from 12 annas (0.75p) to Rs. 8 to Rs. 9/-, the average being Rs. 3/- per acre,[26] the leases fetched immense revenue initially. The state revenue rose from a mere Rs. 2.4 lakhs in 1881-82 to Rs. 4.6 lakhs in 1892-93 recording a total rise of almost 100 per cent in a decade[27] of which land revenue contributed a substantial amount. Rent was increased every year and a large quantity of unaccounted land was discovered. However, a large portion of revenue was lost to the royal treasury due to this leasing procedure, as the land was neither measured nor rentals ascertained. The tenants often withheld the dues when the lessees enhanced the rentals without the king's knowledge. Leased lands too were comparatively small due to the system of settling waste lands on rough estimates without initial survey. The revenue returns were not regularised and the king seemed to remain unpertrubed too, levying various other taxes such as toll taxes, taxes on forest products, duty on cotton exports, ghar chukti tax on shifting cultivator etc. to enhance his treasury.

Owing to a steady influx of the neighbouring erstwhile East Pakistani nationals in search of fertile land, as well as shifting cultivators of the eastern region, population began to rise. The customary rules had to give way to new rules and regulations setting the state's agrarian system awry. Two distinct land use patterns emerged, plough in the plains and shifting in the hills pursued by two different ethnic races plainlanders and tribals, prevailing in two regions. Land could be owned by both the segments of the population unlike several other North-eastern states and this resource being exhaustive could not endure the increasing pressure, hence productivity began to decline in comparison to the population levying a direct physical impact on this typical agrarian economy.

There existed a dualism in the land tenure system of the state owing to the strange geographical situation and the increasing spate of population growth. The king was the sole owner of both the hills and plains, while no rent could be collected from the hills, to enable fair collection of rentals from the plains, three kinds of tenures existed.[28]

(i) A farm (izara) for the collection of rent for a term of years.

(ii) A talook (miyadi) with a rental subject to revision after fixed periods; and

(iii) A permanent talook (Kayemi) the rental of which was fixed forever.

The holders could make their own settlements with the cultivators. The cultivators were lease holders and so did not acquire rights of occupancy. Rents were fair, terms on which leases of waste land were granted were also such as to attract settlers. However, there are evidence that without the king's knowledge the talukdars used to sub-let a part or whole of his holding to one or many under-raiyats at higher rents. These under raiyats were usually cultivators who came across the borders and after harvest often went back even, which meant a loss of revenue again.

During the middle of the 20th century, the land system underwent slight changes. Three classes of talukdars (landholders) remained, though their leases and the rent changed.[29]

(i) Kayemi taluks, revenue of which was fixed in perpetuity,

(ii) Taskhichi taluks, mainly plantations or land granted for special purposes, the rent of which was subjected to revision after 10/15 years, and

(iii) Niskar taluks, which were revenue free estates granted to brahmins as gifts for religious purposes etc.

It is important to point out here that all these systems were applied for the plainland only and catered to the non-tribal population. But with the rise in population of both the segments problems arose regarding settlement of the indigenous people too, who seemed even eager to take up ploughing. Thus tribal reserves were created in 1931 and 1943 for five specified tribes, namely Puran Tripura, Noatia, Jamatia, Riang and Halam. However, these measures were undertaken in much haste and therefore, no proper surveys were made, reserve boundaries were loose and transfers went on unchecked, and so even among the classless autochthons a certain class of talukdars became the owners of vast tracts of land without paying revenue for the same, and simultaneously number of landless cultivators were also on the rise.

Not until 1960 was the Tripura Land Revenue and Land Reform Act (TLR and LR) was passed wherein all old tenures and laws were abolished, and many tenants became actual raiyats.[30] However, despite the government's efforts to create a single predominant class of peasant cultivators in the typically agrarian society a three-tier structure of owner-tenant-owner-cum-tenants features the land tenure system still portraying the inadequacy of the said Act. The peculiarity found in Tripura's case is not the land system evolved but the connection of land availability with the state's population increase, which leaves the land question unanswered. A class struggle thus seems to have ensued since the beginning of the present century—which explains the bearing of land on the development of agrarian economy. The principal elements in this struggle can be pointed out here, viz., the declining ratio of tribal-non-tribal populations alongwith the number of landed properties; the emergence of a new class of tribal/non-tribal rich peasants, money-lenders which was non-existent so far and the alarming rise of poor, marginal peasants and landless labourers. Perhaps, this picture reveals what a devastating impact the land systems, population and governmental policies have had on the economy of the state. One cannot deny that the recent upheaval in the socio-economic-political arena of this once peaceful state is the outburst of the autochthons' pent up feeling of hatred. Ugly provincialism has raised its head in the form of protecting their interests, which naturally has to be at the expense of the other segment of the population. The pre-existing bonds of solidarity among them are not immutable they are historically evolved and as such, are bound to change with the changing times. But the change has not been smooth, not being ushered by the own conscious efforts of those involved, or by identifying themselves as one with the distressed, displaced people which has adversely affected the socio-economic fabric of the state.

Approach to the Problem

It appears from the above analysis land was he only source of sustenance to the bulk of the state's population during the princely regime. Plough cultivation fetched higher returns and so this mode was encouraged while restrictions were imposed on the free access to land for those practising shifting cultivation. The plainlanders were being invited which ultimately proved to be a serious burden as land availability started diminishing. The kings ventured upon tea plantations to utilise the slopes of hills as well as lessen the pressure on land. Thus land was leased out for tea plantations in 1882 for nominal returns, and within a span of a decade forty gardens sprang up, which aimed at employment generating too along with better utilisation of land.[31] Shifting was prohibited here, which ousted hundreds of shifting cultivators from their jobs, and the plantations could not lure them too. Thus coolie labourers came from outside the state adding to the pressure on land in due course of time leading to a scarcity of cultivable land. During this time, 850 sq. miles of fertile land between Langai and Dhaleswari rivers were taken away for campaigning against the Lushais.[32] Demarcation of border took a long time for arbitration and some land were lost in dispute. A considerable extent of fertile land was fleeced off by the British policies reducing the area of the State to half of what it was two hundred years back.

Reduction of the state's geographical territory, leasing out of invaluable hill gradients for tea cultivation, giving away land to the learned men or for religious purposes, shifting cultivation, reserves for autochthons, along with unprecedented population increase led to the present crisis in land availability in the state of Tripura. Ethnicity came to be divided into two categories as immigration increased. Land had to be provided to both simultaneously for livelihood as well as cater to their non-agricultural needs. Therefore, the land factor became a vital issue in the state's economic development. Hence the need of the study Land question in Tripura, in a historical perspective and analysing the impact of the connection between land and population growth seems to be rational in the present context.

Against this backdrop, the study undertakes to depict the various manifestations of the land systems of the last and present century and assess the historical reasons of the amazing rise of population in the state and the devastating impact on its economy. The study also tries to get a clear picture of the nature of land ownership and the different tenures, analyse the various systems of reservations (forests, autonomous districts, rehabilitation projects) and observe their impact on the land question. The study reflects how the economic evils of continuous land transfers, land alienation through covetous means or governmental anomalies have adversely affected the socio-economic fabric of the state. In its concluding part the study suggests a few alternatives, how to utilise land economically and ease the pressure on land as well as cater to the need of the people.

Methodology and Sources

The methodology adopted is basically descriptive and analytical in its narration. It has highlighted the problems relating to the land viz., availability, utility, modes of cultivation, size of holdings, distribution etc. which are the effects of high concentration of population on land. The structural changes in land systems and its economic effects and implications have been studied in an historical perspective keeping in view the interest of the users of land irrespective of ethnicity or class. Discussing these issues in a historical perspective entails a thorough study of the land systems, data on land ownership, rehabilitation projects, reserves for forestry, autonomous bodies etc. during the princely government and those following it. Hence both primary and secondary sources have been made of relevant topics, seminar papers, reports, notifications, promulgations, documents of the princely regime as well as those of the subsequent state governments and various books by distinguished authors consulted for preliminary observations.

The records at the Tripura Secretariat Archives, Agartala and the West Bengal State Archives both at the Writers' Building and College Street have been consulted, confidential letters, notes administrative reports and documents lying at the former have been of great help to get a knowledge of land distribution, land acquisition, as well as various acts and policies on land.

The journals found at the Asiatic Society of Bengal, Calcutta, Annexe Building of National Library, and other historical journals throw some light on the ancient history of the State. Certain epigraphic evidences have been collected from these. Articles in contemporary magazines, journals, newspapers have been consulted which touched upon issues relating to land alienation, land systems, forestry or granting of autonomy in the state.

Materials were collected from several state government offices for information regarding land system of the post independence period. The Census reports gave an idea of the increase of population in the state; Statistics Department provided the records relating to land holdings, agriculture, land allotments, purposes etc. The Refugee Rehabilitation Department was consulted for information of land allotment for refugee rehabilitation purpose and Chakma influx (1980s). The Land Settlement Department came up with several reports/records on the land systems, purpose of land settlement etc. The offices of the Autonomous District Council was consulted for the total area, reserved forests therein, cultivable area falling within its limits etc. Various government gazette notifications, the Tripura Land Revenue and Land Reforms Act (1960) its later amendments, the Forest Conservation Acts, policies were consulted and found useful for the present work.

Issues like transition occurring from shifting cultivation to plough cultivation, jiratia land (border areas cultivated by British Bengal subjects) forest land and land alienation needed personal attention. Therefore, visits were made to Hetalia and West Paikhola villages in Belonia sub-division to get a first hand knowledge of shifting cultivation and the transitionary process. Villages on the borders, Ishanchandranagar, Barpathari Samarendra Chandra Nagar, were visited and information gathered from some original buyers of such land even. Forestry, an absolute necessity to maintain ecological balance also means giving up arable land, so visits were made to Trishna Reserve Forest (in South Tripura) to find out the extent of reservations or opening up of such forests. Plantations, another project of the government to maintain the greenery as well as release pressure on cultivable land were also visited (Rubber Plantations in the Kanchanmala-Sekerkote area). Lastly, land alienation, the inevitable effect of the ongoing land scarcity needed to the studied in details, therefore, visits were made to mixed villages to get a thorough knowledge of the causes of land alienation and its effects. Field data thus collected revealed a lot of currently known and unknown facts. Both published and unpublished books on allied topics on North-East were consulted too to get an idea of the uniqueness of land question of Tripura.

Design of the Study

Broadly the entire work has been divided into nine chapters including the present one which introduces the main theme of the study alongwith a brief history of the state highlighting the crisis in cultivable land in the state. The rationality of the study, methodology adopted and source exploitation have been explained in this chapter.

The second chapter deals with classification of land in Tripura according to its agrarian utilisation pattern. It reveals the tenurial system being followed in the state since the last century and accounts for the fall in man-land ratio. It also envisages a plan as to how most advantageous use of this limited resource can pace up productivity and simultaneously solve the land scarcity problem.

The third chapter presents the grim picture of population increase in the state, its causes and effects. Emphasis has been laid here to show how population pressure on land alongwith enthusiastic rehabilitation schemes, afforestation programmes made available land reach its saturation point leading to misdistribution of land.

Chapter four gives a thorough insight into the rudimentary form of agriculture which is still widespread in the state. The process of transition from shifting to plough is described and its socio-economic implications analysed. The governmental schemes to smoothen the transitionary phase has been examined in this chapter and its pros and cons analysed.

Jiratia land, a peculiar outcome of the princely land system is the subject of discussion of the fifth chapter. The border land of the state used to be cultivated by peasants from across the border who went back to their motherland after independence. A detailed study regarding the extent of this land, their ownership rights, the beneficiaries of this land after they abandoned them and the present utilization of the land has been made.

Chapter six contains a discussion on the role of forestry in the state economy in general and the peasants' economy in particular. The role of forestry has been appraised against the background of the National Forest Policy of the government as it affected the economic development of the area and the population of the state. The continuous process of preservation and destruction have been portrayed which accounts for the dilemma of whether to preserve forests for ecological balance or rehabilitate the scores of people. An answer to dilemma has been attempted by putting forth a suggestion of a judicious policy of afforestation, community management growing of crops and trees together.

Chapter seven deals with the formation of the Autonomous District Council as an attempt to salvage the indigenous masses from economic hardships. A detailed analysis of tribal movements have been made and land identified as the bone of contention. The work done by the Autonomous District Council in this direction has been evaluated critically and the need for a micro-level planning for a faster development of the tribal areas emphasised.

Chapter eight forms the core of the thesis, as land alienation is the ultimate effect of the land crisis discussed so far. This comparatively new phenomenon set-in due to imprudish land system of the yester-years have been discussed in details to detect the causes. The tenancy legislations of princely regime as well as the contemporary land reforms are examined and their shortcomings brought to light. Credit institutions, nexus between money-lender and traders and the legislations enacted to stop alienations are also analysed and their inadequacy in protecting the peasants stressed upon.

The epilogue tries to put forth certain suggestions and recommendations in an attempt towards solving the land scarcity problem. Stress is laid on how to rejuvenate the state economy with limited resources and utilising land to its optimum capacity which can play a significant role in the land question of the state.

Limitation of the Study

The present study encompasses the land question of the state during the 19th and 20th centuries. The main theme of the study i.e. cultivable land scarcity problem has acquired huge magnitude during the second half of 20th century, but to understand the background of this acute crisis one has to delve into the land systems of the yester-years, specially the previous century since both shifting and settled cultivation gained a stronghold along with increase in population in the state. However, a strange dearth of land records or settlements documents of the early 19th century put forth certain obstacles in pursuing the study. Undoubtedly, the archival records do throw some light but these dates from the late 19th century, when the Annual Administration Records of the Political Agency of Hill Tipperah are published, beginning from

1872. Even the first available authentic census data in W.W. Hunter's book "A Statistical Account of Bengal" (1876) is based upon these reports mostly. The kings possessed the hilly eastern part (which happens to be present Tripura) and those inhabiting the hills practised the nomadic tillage called jhum (shifting cultivation) widely, no tenurial pattern could be evolved there of and hence no concrete land system was strictly followed. In the kings Zamindari estate of Chakla Roshanabad comprising of parts of Comilla, Sylhet and Noakhali (now all in Bangladesh) plough was prevalent and the land was farmed out to landlords (talukdars) on leases and the British system was adhered to. The manifestations of the land systems during the said period and its impact on the land of the state is the subject of analysis in the present study. There are several phenomena pertaining to the land question which cropped up in the recent times have been given due weightage. Paucity of records (e.g. Jiratia land) had to be substantiated by personal interviews so the analyst had to depend upon the details furnished by them which may be biased to some extent too. Official records are simply the official version of the events and reveal nothing which may help one to constitute history. These are adhered to and read between the lines to get some knowledge of the subject matter.

NOTES AND REFERENCES

1. *North-Eastern Council: Basic Statistics of North Eastern Region,* 1984, Shillong.
2. *Ibid.*
3. *North Eastern Council: Basic Statistics of North Eastern Region* 1995, Shillong (Data collected by a Task Force on shifting cultivation, Ministry of Agriculture, 1983).
4. *Government of India : Census of India,* 1991, series-1, paper-2 of 1992.
5. Not all land is required by the mankind, only those which can produce crops are utilised. The ratio between such (effective) land area and the total population is known as land-man ratio.
6. The density of population of the different states of the North-East in 1991 are as follows: 10 persons per sq.km in Arunachal Pradesh, 286 persons per sq. km in Assam. 82 persons per sq.km in Manipur, 79 persons per sq. km in Meghalaya, 33 persons per sq. km in Mizoram,

73 persons per sq. km Nagaland and 263 persons per sq. km in Tripura (Census of India 1991, series 1, part-II of 1992).

To take the North-East as a whole, the density of population is more glaring—30 persons per sq. km in 1951 increased to 77 persons per sq. km in 1971 and shot up to 123 persons per sq. km in 1991. This should be judged in the light of the fact that only 30 per cent of the total area of the region is suitable for cultivation and hence productivity only caters for subsistence

It should be remembered that, the density of population implies the average number of persons living in per square km area and not the interpretation of total population divided by total effective land area.

7. Percentage increase of population in Tripura as compared to India during the period 1901 to 1951 (Census Handbooks, Govt. of India, various issues).

8. Annual administration report of the Political Agency, Hill Tipperah 1872 No. 149 dtd. Agartala 31st Oct., 1873. Quoted from Administrative Reports of the Political Agency, Hill Tipperah 1872 Ed. by D.K. Choudhuri Vol. 1—TSCRI, Govt. of Tripura, p. 19.

9. Fleet, J.F.: *Topographical list of the Brihat Samhita;* edited by K.K. Das Gupta, Calcutta, 1973, p. 86.

10. Thomson, W.H.: *Final Report on the Survey and Settlement of Tipperah,* 1915-19, Calcutta, 1920.

11. *Archaeological Survey of India : Copper Plate Charter of Vainya Gupta Gunaighar,* Comilla (507-08 A.D.), Annual Report 1930-34, p. 130.

12. University of Calcutta: *Journal of Ancient Indian History,* Vol. IV, 1970-71, pp. 189-90.

13. Sarkar, D.C. : *Journal of Asiatic Society,* Letters XVII, A copper plate charter by Maharajadhiraj Bhavadeva of Devaparvata, 1951, pp. 83-94.

14. *Archaeological Survey of India : Annual Report,* 1925-26.

15. Majumdar R.C. : History of Ancient Bengal, Calcutta, 1971, p. 8.

16. Das Gupta, K.K. ed : *Journal of Asiatic Society,* Vol. X, 1968, p. 66.

The Sultan agreed and about 1000 families settled at Rangamati (Udaipur) and Ratnapur 500 families at Yasopur and the rest at Hirapur. This event reminds us of the practice of making big collective land grants to the Brahmans and such invasions led to the opening of flood gates of the state for future incursions.

17. Webster, J.E.: *Eastern Bengal District Gazetteer,* "Tippera" Allahabad, 1910, p. 82. We learn that Shah Shuja settled the Udaipur Sarkar (only the plain districts) with Kalyan Manikya for Rs. 99,860/- leaving the hill country.

18. This was Shamser Gazi, ruler of Dakshinsik, who rose from obscurity to power by his ability, Singha, K.C., Rajmala, 1896, pp. 123-126.
19. The king now had to pay Rs. 1,00,001/ as original revenue and Rs. 45,563/ as troops charges alongwith Rs. 1,11,191/ 6 annas 3 paise in thirteen instalments. Khan, A.M. : The Transition in Bengal: Cambridge University Press, 1969, p. 39.
20. On a question of succession the incumbents Ramganga (son of the deceased king) and Durgamani (Yubaraj designated) resorted to the court's decision. In 1809 the Sudder Diwani Adalat gave judgment in the latter's favour, declaring the Zamindari 'an integral portion of an impartiable raj'. The Government gave him the insignia of kingship regarding the hills while the Courts gave him possession of the lands of the plains. Since 1809 every king had to receive an investiture from the British government and required to pay the usual nazar (tribute). A. Mackenzie; A History of Relations of the Government with the Hill Tribes, Calcutta, 1884, p. 274.
21. This is found in the letter of the Secretary to the Govt., F.J. Halliday to the Commissioner of Chittagong, J.G. Harvey No. 121 dtd. 27 Dec., 1838, para 2. Foreign Poll A Dec., 1870 Nos. 362-374, West Bengal State Archives.
22. Report on the Northern Boundary of Tripura State, Yule and Campbell. Settlement of disputed boundary between Sylhet and Hill Tipperah. Letter to the Secretary, Sudder Board of Revenue, Fort William, Dinajpur, 14 Jan., 1851.
23. His Highness Birendra Kishore Manikya's letter to Sir William Marris, Governor of Assam, date not found, Tripura Secretariat Archives.
24. *Imperial Gazetteer of India,* Oxford University Press, London, Vol. XIII, 1905, p. 122.
25. Thomson, W.H.: *op.cit.,* p. 72.
26. *Ibid.,* p. 125.
27. *Imperial Gazetteer of India, op.cit.,* pp. 506-507.
28. *The Annual Administrative Reports of the Political Agency Hill Tepperah,* 1877-78. Prepared by C.W. Bolton Esq. Officiating Political Agent. Dated 29 July, 1878, WBSA.
29. Das, J.N. (ed.): *A Study of the Land Systems of Tripura,* Gauhati Law Research Institute, 1990.
30. *The Tripura Land Revenue and Land Reform Act,* 1960 (No. 43 of 1960) Tripura Government Press, Agartala.
31. Chunta Prakash (Extraordinary Issue) An article by Sri Girija Mohan Samgal—*The Cultivation in Tripura,* 16 Jan., 1930.
32. Aitchinson, C.U.: *A Collection of Treaties, Engagements and Sanads,* Appendix VII, Vol. I, 1892, pp. 77-103.

2

Classification of Land

The surface of earth, comprising of soil, water and other biotic entities, or in other words 'land' has a number of uses and to utilise it properly this segment has to be assessed and subjected to improvement. Land use is particularly dependent upon physiographic, edapho-climatological factors as also socio-economic aspirations and technological know how of a community. Developing land for any particular requirement, without paying heed to its resource potentialities means inviting problems leading to permanent impoverishment of natural resources. Hence land should be classified and further sub-classified in accordance to its 'utility, suitability and capability' for both agricultural and non-agricultural use (viz., forestry, grazing-land, watershed protection, cultivable land, construction sites etc.)[1]. Land classification, initially was based upon physical characteristics, soil character, geomorphological attributes (surface configuration) for their relative permanence.[2] Historically analysing, various ways of land classification experiments were made in the United States of America regarding the geomorphological approach when emphasis was given on regional sub-division[3], regional divisions or homogenous physical conditions, in terms of relief, climate and vegetations were also sought as the basis of land classification.[4] However of late, the concept of classification has under-gone a change and now only three main factors are considered, they are, physical economic and social.[5] Whatever the method may be, to study land question of a state a study of land classification on the basis of its physical characteristics is an absolute necessary. In an agrarian economy, land provides the most important basis for production and also becomes a societal factor. A balance has to be struck between the different users of land and its availability so that benefits can be reaped perpetually. It should be remembered that sustainable development is possible only if the utility of a resource is prolonged so as to permit its exploitation not only by the present but also for the future generations.[6]

As it is, in a typical developing country like ours the balance needs to be maintained between land-use and depletion, and this fact is to be kept in mind while classifying land. The secluded, North-eastern section of the country bounded by Bhutan (West), Tibet (North), China (East), Myanmar (South-east) and Bangladesh (South)[7] depends upon land since industrialisation has not paced up yet. However, most part of this land is highly mountainous and so production of crops is far less than the potentials and chances of bringing more area under cultivation stands bleak in such a situation. The social and economic factors also stand on the way as land-ownership rights are limited to the original inhabitants in most of the states, and they are more adept at shifting than settled farming, which hinders expansion of cultivation considerably. In this chapter an attempt has been made in classifying land according to its utility, capability and suitability with a brief note on the physiography of the state.

The state of Tripura is situated in the southern part of the North-east surrounded by Bangladesh on the West, South and North and Assam and Mizoram in the North-east, lying between-22°56' and 24°32' N and 90°10' and 92°21 E with a land area of 10,491 sq km'[8] of which quite a large area is hilly. Three distinct physiographic zones can be discerned from this landed area, they are, North-South oriented hill ranges, undulating plateau lands and low-lying alluvial plains. The general altitude of the state varies from 780m in the north-eastern part to 15m in the western part above sea-level.[9] A series of hill ranges running parallel to one another (North to South) divides the state into two broad valleys. The principal hill ranges are Jampui, Sakhantalang, Longtarai,

Atharamura and Baramura (East to West). These hills are narrow ridges sometimes so narrow at the top that it is similar to only knife-edged rocks.[10] They are dangerous to walk along and are often impenetrable in the valleys and ravines covered with dense, thick forest outgrowth. There are several low hills too, detached from the main ranges. These flat topped, isolated, low residual mounts (locally known as 'tillas') form an unit in the valley especially round the major plains of the rivers. These are generally used for settlements, plantations or growing kharif crops.[11] The alluvial low lands have developed along the major river beds recently which serve as sites for seasonal cultivation or even rural settlements. Valley side slopes are steep and are of unconsolidated nature due to debris sliding on their walls.[12] These in-filled valleys (locally called lungas) constitute an area fit for seasonal cultivation as they are very fertile and may be further classified into high, medium low and low lunga land, and have low sand content with loamy soil. These may be defined as systematic arrangement of different kinds of land on a virtually permanent basis.

Tripura, during the 19th century was bounded by British districts of Sylhet and Cachar in the North, Chittagong in the South, Tipperah in the West, while its eastern boundary was not well defined. No cadastral survey was conducted in the early times. Assistant Political Agent, U.K. Das reports in 1886-87 that the total area of the state was 4086 sq. miles of which over 2400 sq. miles were traversed by numerous hill ranges and low grounds which were marshy and boggy areas; of the remaining 1600 sq. miles 100 sq. miles were supposed to be actually cultivable and the remaining lay waste.[13] Along the river valleys lay extensive stretches of flat alluvial soil suitable for agriculture.[14] Land was broadly classified into land suitable for cultivation and land not suitable for cultivation. At the time this was feasible because land tenures[15] differed from time to time for the purpose of revenue collections as talukdari rights were given indiscriminately in lieu of service rendered. The plains of Tripura came under permanent settlement in 1793 and later the land tenure system got divided into five distinct classes in 1866. They were, estates permanently settled in 1793, resumed rent free states; chars resumed under

Regulation-II, 1819; 'chars' resumed by means of a civil suit, and estates formerly held by the government but lately sold on public auction.[16] The talukdars (landholders) enjoyed great privilege under royal patronge and in due course, they even became defacto owners of the land, leased out their plots to peasants who tilled them. These tillers too sub-leased their lands to others and in this way, through a process of leasing and sub-leasing holdings passed on to the Korfa tenants i.e. the under-raiyats. However, since they could not keep their stronghold over these fractional plots, they fell easy preys to the whims of the landlords not knowing better utilisation of their land. Collective ownership of jhum lands, gradually made way for private ownerships and here too economic forces operated rendering the jhumiyas landless in a few years' time. With ample land this did not lead to any major crisis but as population pressure increased, the need for a judicious and realistic land use plan had to be envisaged.

Therefore, initially, land survey experiments classified the type of land depending upon geo-morphology and physiographic aspects of land. Later, cropping pattern deciphered the 'type' of land and so got further 'classified'. Later still, the topography, nature of the soil, need of the time etc. were considered to augment state income and put land into economic use. The upland (tilla) soil, a reddish, sandy loam type soil was found to be deficient in nutrients like phosphorus, nitrogen and potassium perhaps due to its continuous exposure to the tropical sun and torrential rains.[17] This rendered the soil to be devoid of the essential ingredients required for crop production. Only horticultural and forest plantations can be found here other than shifting, the vegetation is thin with shrubs and uneconomic outgrowth. This type of soil covers almost half the total geographical area of the state. The midlands, which constitute up to 15° land slope are generally more fertile and sugarcane, mesta, ground nut, chillies and other kharif crops can be grown abundantly. In the river basins or alluvial plains, the soil is sandy clay loam to clayey and is suitable for paddy, jute, potato, mustard and vegetables, hence, are most resourceful. The whole state has been divided into five main land-use classes for sustainable use of a parcel of land by the Land use Board.[18] They are built up land,

land for agriculture, forests, water bodies and others. To facilitate cultivation method these classifications are further categorised on the basis of topography, environmental and edaphic factors and also the economic requirements.

On analysing the present land use classification in Tripura, as shown in Table 2.1, one may notice that quite a considerable land area is taken up by forests and the North district particularly has the largest forest area, yet there has been quite a loss and degradation of forest cover which calls for providing appropriate protection to be meted out. The uncultivated land is the least in west district and this is because this region has the largest river valleys and is most densely populated and hence intensively cultivated too, where as uncultivated land is abandant in the Northern district of the state owing to its hilly topography. Due to rapid urbanisation in the west district land not available for cultivation is the highest but at the same time the net area sown is highest too owing to intensive cultivation pattern being followed here, as area sown more than once accounts for 86,200 hectares almost double that of the other two districts. Even as land available for cultivation is lowest in the Northern district, the groves, orchards take up quite a large chunk of land (22,840 hectares) an indication towards optimum use of land. However, one should remember that such classification merely are official and suffices the general demand for the time being as land is the most dynamic natural resource and so cannot be classified into water-tight categories. The utility of land depends upon the need of the hour which can be explained by the following example of area under shifting cultivation. The indigenous peasants often fall back upon shifting mode of agriculture with increasing pressure on land. It was found that in 1975 the area under shifting was only 2.6 per cent of the total area but within a period of 24 months it increased to 10.76 per cent in 1977.[19] As a consequence of such use most of the hill slopes where jhum takes place is susceptible to erosion and with the decrease of fallow periods, turn barren or is covered by thin forest vegetation. Obviously, such changes owe its beginning to the abnormal increase in population of the state, who often occupy low lands. Over dependence on shifting usually decreases crop production too due to loss to soil fertility. However, these statistics of land indicates that there is still vast scope for bringing more cultivable waste land under plough, specially in the North district. This needs to be taken care of because leaving land utilized for long periods often lead to permanent degradation hence efforts are to be made to put them to better use e.g. forestry, suitable agriculture use so that further degradation can be arrested.

Table 2.1: Land—Use Classification in Tripura

(Area in hectares)

		West District	*North District*	*South District*	*Total*
1.	Geographical area	299682	387386	361501	1049169
2.	Area under forest	105073	255398	245697	606168
3.	Land put to non-agriculture use	61106	35812	36582	133500
4.	Barren uncultivable land	N.A.	N.A.	N.A.	N.A.
5.	Land not available for cultivation (3+4)	61106	35812	36582	133500
6.	Permanent pasture and other grazing land	N.A.	N.A.	N.A.	N.A.
7.	Land under Misc. tree crops, groves not-included in net area sown	1080	22840	3080	27000
8.	Cultivable waste	180	230	190	600
9.	Other uncultivated land excluding fallow (6+7+8)	1260	23070	3270	27600
10.	Current fallow land	2093	1456	1002	4551
11.	Other fallow land	150	300	250	700
12.	Total fallow land (10+11)	2243	1756	1252	5251
13.	Net area sown	130000	71950	74700	276650
14.	Area sown more than once	86200	47650	49500	183350
15.	Total Cropped area	216200	119600	124200	460000
16.	Cropping intensity	166	166	166	166

Source: Statistical Abstract, Tripura, 1994 Directorate of Statistics, Government of Tripura, pp. 101-102.

* During 1993-94. ** N.A. = Not Available.

Land classification, here tends to be along cropping pattern subjected to the dynamism of land-use. Therefore, we find highlands are single cropped areas, where as multiple-cropping is practised in the valleys. Rubber plantations, tea plantations too take up gentle slopes or plains. In the past large chunks of forest areas were completely deforested while trying to provide settlements, industrial areas or even economic security other than wanton felling of trees for timber, fuelwood etc. However, this has been made good and in 1991 census forest area has increased considerably. Yet one problem remains—increasing population and decreasing cultivable land. The total irrigated area in the state is meagre (about 10 per cent). This paucity of agricultural land along with terrain constraints emphasise the need for adopting scientific land-use planning to achieve maximum benefits without disturbing the ecological balance. Land, the most important resource of the state is the most scarce resource too. Shifting cultivation renders acres of land uncultivable subjected to soil erosion, a most undesirable factor, Valuable forest resources have been depleted. Yet the physiographical characteristic of the state clearly indicates that forestry, besides agricultural cash crops will be able to serve as the basis for economic livelihood. It may be said that the forests can achieve socio-economic development of the state.

Urban use of land is another alternative resource factor. With the development activities, land use pattern changes along with its ownership pattern i.e. from agricultural use to residential settlement-oriented or industrial use.[20] It is noticed that a piece of settlement land is of greater value than a piece or agricultural holding. With urbanisation growing, land is required to be classified for settlement purpose too so as not to hamper the cultivable plots. Perhaps, in Tripura's case, this is not as widespread as in other parts of the country save for the capital Agartala where land prices soared high as pressure on urban land mounted. Yet since the problem persists, this needs to be regulated by the legal framework of a comprehensive reform policy. It has come to the notice that concomitant land hunger satiated by individual property owners extinguishes the rural i.e. agricultural land use system.

In an agrarian economy like Tripura, more than 84 per cent of the population live in the rural areas and in some way or other are engaged in cultivation. But land per person i.e. land man ratio (1991 census) is 0.38 ha. only—which is the lowest in the north-eastern region, except Assam. The agricultural census (1985-86)[21] reveals that most peasants hold land between 0.02 and 0.5 ha. and even the data of 1990-91 shows an increase of such holdings by 2.61 per cent and a decrease in the operated area by 18.48 per cent. (Basic statistics, 1995). Such small holdings are bound to be improperly used and the users are left with a meagre income. Intensive irrigation, fertilizers, better technical know-how may increase productivity of such plots, but in Tripura's case such possibilities are bleak due to lack of finances. The Agriculture Finance Corporation in their report (1983) has suggested that even unirrigable tilla land may be brought under production by proper land use planning.

Thus classification of land is to be undertaken in such a way so at it escalates production. Due to certain bottlenecks in administrative levels, the actual users of land, very often do not or cannot acquire land. Here the need of a regular scientific land classification arises to regulate the economic use of land, since it provides the most important source of sustenance in the absence of industrial infra-structure. The steps taken in this direction is to take care of the bonds between the user and land, specially in case of the tribesmen. The concept of land reforms requires to be widened and not be limited to the general idea of redistribution of land or fixing of land ceilings. Rather, the new policy needs to be formulated on the lines of land capability classification categories like requisite land area for crop cultivation, area to be brought under cultivation, area of reclaimable land, pattern of ownership and land for developmental projects.

The land question of Tripura needs to be aptly answered by way of classifying land according to necessities. In the face of the abnormal demographic revolution and demand of land from various non-agricultural sectors e.g. ONGC projects, expansion of administrative office complexes, district offices, schools, institutions, human settlements etc. classifying land according to its capability will perhaps be able to solve the present land crisis of the state. There has been drastic changes in the land-use pattern of Tripura since the last century, moreover, shifting cultivation has rendered miles and miles of land infertile. The forest lands also have had to be opened up for various necessities, these often lead to mass destruction of land in the future. The commitment of land to an undesirable use would lead to its degradation, thus making it unsuitable for that particular use later. It may be pointed out that short term benefits have adverse effects on land in the long run.[22] Therefore, in view of increasing demand for food production, high priority is to be given to the promotion of optimum land use; as also its conservation. Taking into consideration the various needs of the society and its technological possibilities the basic needs of the society should be envisaged first, because as the economic interests get diversified, the capacity of the society to feed itself also decreases. Not only that, giving away of degraded marginal plots of land to plough leads to further degradation and eventually results in land loss—which is obviously not desirable in Tripura's case.

NOTES AND REFERENCES

1. De, N.K.: *Measuring Land Potentials in Developing Countries;* University of Burdwan, 1984, p. 2.
2. Powell, J.W.: *The Physiography of the United States,* American Book Co. New York, 1896, p. 345.
3. Davis, W.M.: *Geographical Essays,* New York, Reprint 1954, p. 772.
4. Joerg, H.L. : *The Sub-division of North America into natural regions—a preliminary enquiry Annuals of AAG,* Washington D.C., 1914, Vol. IV, pp. 55-83.
5. Kellog, C.E.: *A Method of Rural land Classification,* U.S.D.A. Technical Bulletin No. 169, 1935.
6. Land use and conservation Board, Dept. of Agriculture & Co-operation, Ministry of Agriculture, New Delhi, 1993. A discussion paper.
7. Centre for study of Man and Environment, Presidency College, Calcutta (Prepared by A.B. Biswas, A.K. Saha, S.P. Das Gupta and C. Chakraborti): *State of Environment in Tripura,* Dept. of Science, Technology & Environment, Govt. of Tripura, pp. 5-6.
8. *Census of India,* 1991 series—1 paper 2 of 1992.

9. Saha, Das Gupta, Biswas, Chakraborty: *op.cit.*, p. 5.
10. Capt. Badgley (Dy. Supdt. of the Topographic Survey): *Narrative Report of the Hill Tipperah, North Chittagong and Lushai Hills*—A Topographical Survey, 1872-73.
11. *Research Bulletin No.* 23 pp. 18-19.
12. Saha, Biswas *et.al., op.cit.* pp. 10-11.
13. *Annual Administrative Report* (1886-87). Prepared by Assistant Political Agent Uma Kanta Das (Ex-officio Political Agent, Mr. J.C. Price) July 7, 1887. (WBSA).
14. Bose, Asoke (Esqr.): Notes on the Geology and Economic resources of the Southern area of Hill Tippera, Geological survey of Hill Tipperah. (As submitted to the Durbar for consideration)—1909 A.D.—Tripura Administrative Records.
15. Land tenure here identifies the ownership of land and the relation of the owner with the cultivator. During the earlier times tenurial system were simply by produces of various kinds of settlements made from time to time.
16. Hunter, W.W.: *Statistical Account of Bengal,* Vol. VI, pp. 401.
17. Gautam N.C. and Channiah: *Classification of Landsat data using MDAS.* Land use Board Bulletin, 1993 Agt. Tripura.
18. Research Bulletin No. 23, *op.cit.*
19. Land use and land cover change (Percentage)—Annual perspective plan, 1993 Land use Board, Agartala.
20. Som, P.K.: *Land use oriented Land Reform Policy* (in Manjula Bose ed. Land Reforms in Eastern India, Calcutta.
21. State Report on Agricultural census, Tripura (1985-86) Vol. II, Agricultural census Commissioner and Directorate of Land Records & Settlement.
22. National Consultation on perspective plan for conservation. Management and Development of Land Resources of the country. A discussion paper Oct. 21-24, 1991 New Delhi.

3

Population Pressure on Land

The study of population is important from the point of economic welfare of a region. Human beings are both the means and ends for development and are considered to be asset as well as liability. Human resource is an asset when it contributes to the labour force, market demand and even takes an active role in developmental work. However, it may be a liability too if the number increases beyond the economy's capacity to sustain. In the present chapter, the aim is to decipher whether the state's population growth is an asset or a liability, hence the knowledge of the size of population, rate of growth and ethnic composition is attempted and the causes and affects there of have been analysed.

The North-eastern sector of the country constitutes approximately 8 per cent of her total geographical area and its population forms 3.7 per cent of the country's total strength.[1] The population growth here is more a result of mass-scale immigration than natural, which accounts for the rapid increase in comparison to all-India figures. Immigrations had started in this part of the country since time immemorial. The original inhabitants of this region belonged to the Indo-Mongoloid race who moved here from the northern region of the country in search of arable land around 2000 B.C.[2] They belonged to the Indo-Chinese linguistic family i.e. Tibeto-Burman and Mon Khamer group. Reaching the fertile valleys of Assam, they decided to make it their permanent abode. But since they practised shifting cultivation, they could not settle in one place, though they slowed down their shifting and roamed the neighbouring gentle slopes in frequent intervals. Subsequent waves of immigrants settled in other North-eastern states some even went across to Arakan, Burma and Chittagong.[3] Therefore, the immigrants kept on swelling the population of the region. Yet it must be added that of all the states, the immigration into Assam and Tripura surpassed all the known records of migration in modern times. This manifests itself in the growth rate of the state population according to Census Records which shows a rise of 137.80 per cent during 1901 and 1951 in Assam and 273.41 per cent in Tripura during the period while that of India grew by 51.50 per cent. The influx in Assam may be contributed to the mass scale immigration of tea plantation labourers added to the waves of frequent migrations from East Pakistan. In Tripura, the beginning of the influx can be traced back of the "invitations" meted out by the kings to the plainlanders. Tripura, unlike the other North-eastern states which are mostly hilly, possesses certain plain lands too. But these invitations soon paved the way for mass scale influx following communal upheavals and this changed the demographic pattern of the state. Initially such immigrations did lead to the strengthening of the regional economy as the total cropped area increased, the agricultural production shot up and even vast areas of waste land was reclaimed. But the continuous inflow of new entrants soon adversely affected the regional economy as land was the only means of sustenance, and the gradually increasing population pressure created acute economic problems for the future—low man—land ratio, fragmentation of holdings, low productivity, social unrest etc.

Tripura has always maintained a high growth rate as an aftermath of the royal policies and this is revealed in Table 3.1 when the growth of population is compared to the region and the country as a whole. The Table reveals that the growth rate of the North-eastern region is always higher than the all-India average of which Tripura contributes quite a huge number. During the decade 1911-21 India experienced a negative growth rate (-0.30) but the population of North-east rose by 18.60 per cent while that of Tripura rose by 32.59 per cent. The decade 1951-61 showed a tremendous growth in Tripura's population (78.71 per cent) while that of the country rose by 21.64 per cent only. This fast rising population trend during this century may be attributed to slow falling birth-rate and declining death rate but most important is the fact that a majority is contributed by mass influx of displaced people from neighbouring countries or inflow of people from other parts of the country in search of a livelihood. This alarming rate of population growth slow down the pace of development of the region, cannot absorb the population in the labour force for productive activities leading to a stagnant economic set-up. There are no dearth of natural resources or investments but the development potential of the region shows no sign of picking up thus such abnormal growth rate in this region obviously leads to much adversities. The reason of high population growth in the North-east cannot be attributed to simply high birth-rate and low death-rate, but it lies elsewhere which is to be analysed in the present chapter.

Table 3.1: Growth of Population during the Decades of 20th Century

(in percentage)

Year	*Tripura*	*North East*	*India*
1901-11	32.48	18.40	5.73
1911-21	32.59	18.60	-0.30
1921-31	25.63	19.40	11.00
1931-41	34.14	20.10	14.23
1941-51	25.86	19.10	13.31
1951-61	78.71	41.20	21.64
1961-71	36.28	35.00	24.80
1971-81	31.94	35.90	25.06
1981-91	34.30	NA	23.85

Source : Census of India, compiled from various issues.
NA : Not available.

Population Growth in the State: A Profile

Geographically, this tiny border state of Tripura shares more than three-fourth of her boundary with Bangladesh. More specifically, the western part of the state corresponds to the eastern fringe of Bangladesh and low altitudes, ridges or narrow rivers do not act as barriers or even natural impediments to migratory movements. The Indo-Mongoloid races which migrated from North India in pursuit of fertile land reached Tripura in due course. This accounts for the fact that the indigeneous people of the state (viz. Tripuris, Reangs, Halams etc.) bear an ethnic affinity towards the Tibeto-Burman group of the Nagas, Bodos, Kukis inhabiting the adjacent states. The earliest reports, those of the Political Agent Mr. A.W.B. Power[4] opines that even in 1872 the entire population of the state could be divided into two distinct categories—the inhabitants of the hills and those of the plains. The former, according to the Political Agent consisted of six major tribes of Rajbanshis, Tipperahs, Jamatiyas, Noatias, Reangs, Halams, and Kukis while the latter were mostly Bengalis and Manipuris. Both of these categories together were approximately 35,000 in numbers (20,000 hill tribes and 14,500 plainlanders). Around the same time 4000 Chakma families came to Tripura in pursuit of jhum land.[5] Initially, they settled in the upper waters of the Gomati river, those lush green slopes which were abandoned by the indigeneous shifting cultivators due to recurring Lushai raids in these region. The coming of the Chakmas was considered to be a boon by the king of the state and so they were encouraged to settle on vast tracts suitable to their needs. The Chakma influx[6] continued for quite some time during the last quarter of 19th century. They turned towards Tripura due to several factors, like population growth, dearth of arable land, being forced to serve as coolie labourers for the Lushai campaigns much to their abhorrence. The Mogs[7] whose original home, like the Chakmas, was in the Chittagong Hill tracts, too moved into Tripura during 1882-83 to live here permanently, occupying vast expanses of agricultural land in the southern part of the state i.e. Belonia and Sabroom. Internal strifes, search for jhum land, population growth are supposed to have made them turn towards Tripura. The Kukis, a war-like tribe of the same ethnicity as Lushais, speaking the same language too got themselves enlisted in the royal army proving their prowess, and serving the king loyally, stayed on. All these tribal immigration obviously added to the local population figures which hiked the total numbers thus inflating the population of the state.

Other than the immigration of the same ethnic group. There have been immigration of other tribal people from Assam, namely the Garos, the Assamese, the Bodos and also some Khasis during the early decades of 20th century.[8] They came mainly for their agricultural pursuits—the Garos searched for jhum land, the Khasis wanted to grow betel leaves in Dharmanagar region. Besides these, tribes from other parts of India, with no ethnic affinity to the local inhabitants also reached Tripura. The first tea estate was started in 1916 in Kailasahar division of the state. Need for coolie labour rose since the autochthons were quite unwilling to take up such arduous jobs. Hence, coolie labourers were called from states of eastern India, Bihar, Orissa, Madhya Pradesh and tribes namely Mundas, Oraons, Bhils, Santhals, Bhutias, Lepchas etc. immigrated to Tripura. Moreover, once developmental projects were gaining importance, brick kilns, factories, road-building, other constructions were required. Many daily labourers came over from Assam and other neighbouring states attracted by the prospects of better living. A perusal of Table 3.2 will given us an idea of the total tribal immigrant population of the state since the last quarter of the 19th century till the 1931 census. Table 3.2 shows that quite a large number of tribal immigrants came and settled in Tripura in search of agricultural land related jobs. The neighbouring tribals tops the list of immigrants while those from Assam or Chota Nagpur came as labourers. Naturally, in due course of time, they acquired some land and settled down adding to the pressure on land.

However, the other segment of the state population i.e. the plainlanders far outnumbered the tribal immigrants. They comprised mainly of neighbouring Manipuris and East Pakistani immigrants–both Hindus and Muslims. The Manipuris had renounced shifting mode of cultivation altogether for plough and inhabited the valleys of the people of the plains. Affable, pleasant and quite thriving these Manipuris immigrated from the Sylhet district. This immigration was perhaps an outcome of the goodwill gesture arising out of the matrimonial alliance between the two royal families.[9] Foreign aggressions (Burmese wars) revolution in the state, economic forces too acted as an important factor in their influx. They preferred living near the forest peripheries, reclaiming land for agricultural purposes, even sold land to immigrant Bengali cultivators.[10] The easy terms offered by the kings seemed to act as an important inducement too, and they settled in large number in Agartala, Khowai, Kailashahar and Dharmanagar divisions numbering about 19,210 (1931 census).[11]

Table 3.2: Tribal Immigrants to Tripura, 1931 Census

	Name of Tribe	*Hailing from*	*Population*	*Occupation*
1.	Chakmas	Chittagong Hill Tracts	8613	Jhuming, plough cultivation
2.	Mogs	-do-	5687	-do-
3.	Garo	Assam Hills	2740	-do-
4.	Lushai	-do-	2000	-do-
5.	Bodo	-do-	181	-do-
6.	Khasi	-do-	23	-do-
7.	Oraon	Chota Nagpur	979	Tea Garden Labourers
8.	Kanda	Orissa	667	-do-
9.	Kurmi	Bihar	338	Agriculture
10.	Munda	Chota Nagpur	2058	Tea garden labourers agriculture
11.	Santhal	Santal Pargana	735	-do-

Source: S.C. Deb Barma, Census Report, (1931 A.D.) Agartala, 1933, pp. 51-57.

The Manipuris were outnumbered by the immigrants from erstwhile East Pakistan in no time and this immigration surpassed all others and gradually posed a colossus burden on the state. Initially, vast expanses of arable land remained under dense jungle consisting of bamboo, cane, grass etc. which made the interiors inaccessible for half the year. As the Manikya kings ushered in a good government, they felt the need of able educated administrators and thus embarked upon a novel process of "inviting" such people from the neighbourhood regions. In so doing it was found that the king of the state, although owner of the entire state, holds comparatively little land in his own hands, the greater portion being given to the talukdars and farmers. As the educated elite from the neighbouring regions, came here, they needed land to settle down, moreover, during the end of the last century there was a growing intensity of educated unemployment in the districts of Bengal.[12] As they began to settle here the kings seemed to gain doubly, since these people, though not essentially peasants, started expanding plough cultivation which increased the revenue returns of the king at the same time, they helped the king in the administrative reforms. Imperial Gazetteer records an increase of state revenue from 2.4 lakhs (1881-82) to Rs. 4.6 lakhs (1892-93) i.e. nearly a hundred per cent increase.[13] The growth maintained this rate even in 1903-04 when the total revenue amounted to Rs. 8.17 lakhs of which land revenue alone amounted to Rs. 2.32 lakhs and this was entirely from the holders of agricultural land in the plains. The average rate of rent at the time worked out to be Rs. 3 per acre. On this basis it is estimated that about 30,000 hectares of land was already settled for agricultural and homestead purposes by that time, which once again shows that by the end of the 19th century immigration of cultivating peasants from the British-Bengal districts took place in large numbers.

To substantiate the aforesaid view it may be stated here that high caste Hindus, hard working Muslims, the agriculturists were induced by the kings to settle in the state on easy rental terms. As ownership rights could not be acquired easily in the state, till the end of the 19th century (1870s) a tenancy Act was passed in 1885-86,[14] under which British Bengal subjects were given occupancy rights on their land, the rates of which were kept low to attract the peasants from elsewhere. The process of "invitation" already had added quite a sizeable number to the population of the state. The departments of judiciary, forestry, education (almost all the teachers were Bengali plainlanders) were filled up by the immigrants. The menial jobseekers like the washerman, barber, mali, patni found Tripura to be a more attractive place to settle in than the British districts of Bengal. This immigration was further helped at that time with the aim of bringing about development in that area, the system of granting Kayemi taluk was introduced in

Kailashahar district.[15] Many reclaimed land under jungle for cultivation purpose. The famines (1865-66)[16] too prompted many to seek refuge in the state. Thus a steady inflow of the neighbouring plainlanders, irrespective of caste, creed or class led to a considerable hike in state income along with expansion of cultivable land, though there was hardly any inflow during the end of the previous century, perhaps their conservatism hindered their migratory movements at that time. But as certain abnormal conditions, such as natural calamities political unrests cropped up the farmers were driven out of their homeland to new places where lay prospects of acquiring new land. It is found from authentic sources that the tenure of land holding under the zamindars was quite insecure. In the Ryotwoari areas, where the peasants (ryots) came directly under the government the latter became the 'oppressors' in respect of levying high rate of land revenue on the peasants.[17] Bipan chandra brings out the fact[18] as follows—"Heavy assessment of land was one of the main causes of the growth of poverty and the deterioration of agriculture in 19th century"; even Bishop Heber (1826) wrote "—the peasantry in the company's provinces are on the whole worse off, poorer and more dispirited than the subjects of the Native provinces—the fact is, no native prince demands the rent which we do."[19] Droughts, famines, pestilence too forced the agriculturists to leave their birth places in search of a living in the neighbouring states.

Thus, the 'pressure' on land continued but by no means could the numbers be ascertained, and by the end of the 19th century a considerable portion of the available land in the plains was lost to the state. Land grants were being made at random to high caste Hindus, Muslim pirs, fakirs, even high officials (Vaidya-kayastha) by the kings. Nearly a fourth of the total geographical area was given away as "Devattara" or "Brahmottara" lands. Certain evidences can be cited here:

(a) Satadronbhumi, a jungle land in the Udaipur mouja was given to Sri Mukunda Bhattacharjee free of rent to be enjoyed freely (1573 Saka era).[20]

(b) The priest of Tripureshwari Temple—Ratnanarayan was granted certain rent-free land.[21]

(c) Rent free land measuring 8 drones and 4 kani was given as an award to servants Harinarayan Rambansh, Sitaram Choudhury (1171 T.E.).[22]

(d) In Haripur mouja, a half jungle and half arable land was given to Haridev Pandit (1086 T.E.)[23]

The Muslims too received government patronage as there is evidence of a growing Muslim immigrant population. The local tribal people preferred shifting cultivation even though plough was not altogether unknown to them. They occassionally employed Bengali Muslim cultivators as servants who tilled some waste land belonging to the former. Muslims immigrated to Tripura mostly for seeking agricultural land.[24] They were more hard-working and often made their way to the interiors of the state as an aftermath of several earlier inroads many even stayed over. Even the Political Agent, A.W.B. Power in his report of 1872 gives the figures of Bengali population, chiefly comprising of Muslims as 14,500 who settled mostly in the southern districts of Udaipur, Sonamura and Kailashahar in the North outnumbering people of other religious. Even the kings patronised them by giving away rent free land—

(a) Arable land in Fulpur mouja was given to Sri Khandakar Abdul Gani during a festival in 1578 saka era.[25]

(b) 5 drones of land was granted to Kaji Mumin, Mansur, Hussain for religious purposes in Udaipur and Meherkul mouja, which could be enjoyed hereditarily[26].

With such a growing affinity between the two states, it was found the their numbers grew continuously. Whenever a relative, a friend or an acquaintance seemed to be in distress, those already settled held out a helping hand and thus population grew by leaps and bounds unheeded as seen from Table 3.3. The Table reveals the increasing population pressure on the land available. The earliest Census report (1874-75) shows a large number of Muslim people residing within the state. This proves the fact that not only did the king give away land to Muslim kazis and Moulavis, even large number of peasants arrived here with the hope of acquiring some land and making it cultivable with their hard work. But within

Table 3.3: Religion—wise distribution of Population

	Hindus	*Muslims*	*Buddhists*	*Christians*	*Sikhs*	*Jains*	*Others*	*Total*
1874-75	4,339	14,228	–	113	–	–		74,523
1875-76	6,256	18,612	–	133	–	–		1,37,575
1901	1,19,192	45,323	5,999	137	–	–		1,73,325
1911	1,58,101	64,953	5,997	138	–	–		2,29,613
1921	2,07,696	82,288	10,149	1,860	–	–		3,04,437
1931	2,61,589	1,03,720	14,351	2,596	14	–		3,82,450
1941	3,47,792	1,23,570	7,724	328	–	–		5,13,010
1951	4,80,662	1,36,940	15,403	5,262	–	–	741	6,45,707
1961	8,67,998	2,30,002	33,716	10,039	195	49	201	11,42,005
1971	13,93,689	1,03,962	42,285	15,713	375	318	375	15,56,342
1981	18,34,218	1,38,529	54,806	24,872	285	297	227	20,53,058
1991	23,84,934	1,96,495	1,28,260	46,472	740	301	2(-)	27,57,205

Source : (1) Administrative Reports of Political Agents, Vol. I (ed) by D.K. Choudhuri, Appendix M, pp. 94-95.

(2) S.C. Deb Barma : Census of India (compiled) pp. 26-29.

(3) Census publication, Various Issues.

the turn of the century their numbers fell behind the Hindus. Hence Hindus constituted the Bengali (Hindu) plainlanders Manipuris, Nepalis, Marwaris and others who came here as well as it attributes to the fact that the hill tribes were gradually getting converted to Hinduism during the early years of the 20th century. However, the immigration into this state during the next thirty years (1931-61) changed the entire profile of the state beyond recognition.

The partition also led to sacrificing of 70 acres of cultivable land under the Chakla Roshanabad (East Pakistan) and the cultivators over there who were mostly the Tripura king's subjects made a bee—line for the state being evicted from their own land. They were accommodated at the King's insistence. The communal strifes in the years preceeding Independence (1930-47) in Dhaka, Noa-Khali, the persecution of the Chakmas of Chittagong Hill Tracts were particularly responsible for this change. A relief and rehabilitation cell was set up under the Revenue department of the royal government which was later converted into a separate department after the Noakhali riots (1946). The atrocities continued well into 1950s and the department was developed into the Directorate of Rehabilitation after Tripura's merger with the Indian Union (1949). During this time the trend of population of the state definitely showed an increase by leaps and bounds and even had an excess over India's population.

Table 3.4 portrays that the decade 1931-1941 shows a tremendous increase in population, as high as 34.14 per cent (or 1,30,560 people in absolute terms) a proof of heavy immigration. World-wide depression affected the jute growers of British Bengal, reducing their earnings considerably, leading to mass unemployment and economic distress. A large number of agriculturists, middle-class people came in search of jobs or any other means of livelihood to this neighbouring state. The next decade, 1941-51 shows a slowing down of immigrant population as the rise in percentage fell to 24.56 per cent (or 1,26,011 people in absolute terms). This trend can be analysed thus—influx slowed down after the riots were over, several sought refuge in other parts of India and most important fact, following the communal atrocities there was a mass scale Muslim population exodus to East Pakistan the

Census reports show a minimum rise in the Muslim population from 1,03,720 to 1,23,570. Perhaps being aware of the political-religious adversities they deterred from coming to Tripura. However, the following decade, 1951-61 recorded a huge increase of 78.71 per cent (in absolute therms 5,02,976 people). This may be explained that the Directorate of Rehabilitation was set up in 1949 the refugees were 'settled' and their settlements recorded. Also the communal atrocities forced thousands to leave East Pakistan and seek refuge in an amicable region. However, it should be mentioned here that a great number of Muslims decided to cross over to East Pakistan and the Table 3.3 shows a remarkable drop in their numbers after 1961.

Table 3.4: Trend of Population of Tripura since 1931

Year	*Population (Tripura) in lakhs*	*Population (India) in lakhs*	*Percentage excess over India*
1931	3.13	1797	14.63
1941	5.13	3185	19.91
1951	6.46	3610	12.55
1961	11.42	2164	57.17
1971	15.56	5479	11.48
1981	20.53	6852	6.88
1991	27.57	8463	—

Source : Some basic statistics of Tripura (compiled from various Census Reports).

The rate of influx did not however stop here rather, gathered momentum once again during the Bangladesh liberation war (1971). Fear of atrocities as well as victimisation led thousands to flee across the border into a more hospitable country. The census reports (Table 3.1) show a marked increase of 36.28 per cent or 4,14,337 in absolute terms though lesser than the previous decade but by no means did the momentum decline. One should bear in mind that the saturation point was reached by this time and land could not provide sustenance to the scores of people who still trudged in seeking refuge. They now began to 'settle' by covetous means ousting the locals even. This trend continued through 1981 when the population rose by 31.92 per cent (4,96,716) and in 1991 by approximately 34.30 per cent (7,04,842). It may be noticed here that since the last three decades there has been a distinct rise but in a subdued tone, however, which reveals the fact that not all who 'settle' are recorded. The chief feature of this hike in population trend is that the non-tribal population almost doubled up during the past three decades, 7.82 lakhs in 1961 to 11.05 lakhs in 1971 to 14.69 lakhs in 1981 census, which explains the nature of population explosion attributed to immigration.

Another feature of population explosion may be cited here, the Chakma influx. During the late 19th century, Chakmas were already entering the state in search of land for shifting cultivation. The installation of Karnafuli Dam project[27] in 1940s displaced a number of Chakmas from their native land. After the partition, the Chittagong Hill Tracts were declared 'Excluded Area' and later included in Pakistan (17 Aug., 1947). This inclusion accentuated certain socio-political tensions in this area mainly because the inhabitants were ethnically and religion wise different too. Muslim plainlanders were being settled in the hill tracts which made the Chakmas of the region minorities in no time (97 per cent in 1947 to 55 per cent in 1985)[28]. Religious persecution, ethnic violence which had its roots in economic reasons forced the Chakmas to enter India en masse, their nearest refuge being the states Mizoram and Tripura, 50,000 Chakma refugees entered Tripura in April, 1986 and were accommodated in six relief camps of Lebacherra, Takumbari, Pancharampara, Korbook, Shilachari and Kathalcherra, under Amarpur sub-division, by July, 1989 their number rose to 67,000.[29] However, when they had to return to Bangladesh following mutual understanding between the governments there were only 53,418 refugees recorded, which meant they had assimilated with the local population and were reluctant to go back, fearing a repetition of the persecutions, unfavourable living conditions and even no regular income to survive upon.[30]

This unprecedented increase in the population has made trans-border migrations complex and controversial as the pressure of population on Tripura's meagre economic resources rose. Most of these displaced people sought settlement on land for their livelihood and the government took upon their shoulders the task of settling them. Even forests, low hillocks, waste lands, swamps were given away indiscriminately. It is found that of a jote land at Hatileta[31] about 12.80 acres were allotted to few refugee families who cleared dense forests and reclaimed land for settled cultivation. 16 drones or 102.40 acres of plain land and 9 drones, 8.5 gandas of tilla land were given to refugees in the Rangutia mouja[32] (Bamutia tehsil) for plough cultivation. These lands belonged to certain local inhabitants, but were lying waste as the government put it and so were given away to the refugees so that it would be put into economic use, without the permission of those to whom it belonged, neither were they compensated.[33] However, the Refugee Rehabilitation department claimed that only disputed and undocumented land were being given away, and since the acquisition process was lengthy and settlement was to be given immediately, rehabilitation was done in a haste. Thus, land became the sole source of income for the scores of people who flocked to the state as displaced persons, and being adept at proper land utilisation, were pampered by the Government too. In fact the over-zealous policy of the state government, after partition, evicted several original owners without any prior notice or even due compensation. By 1961 there resulted quite a vast expanse of uncultivated, fallow lands (half-million acres) which were grazing lands mostly, "much overgrazed and eroded . . .",[34] which left little or no scope for expanding agriculture. Pressure of population already had a bearing on the land of the state. It may be inferred that owing to strange geographical contiguity, governmental policies, abnormal politico-religious upheavals, economic compulsions in the neighbouring regions, the population of the state kept on rising. The hike in the initial years did have a positive impact as the cultivated area increased pushing up the state (land) revenue along side, which could be termed as an 'asset' undoubtedly. The main income of the princely state till now revolved around toll taxes and family taxes and land revenue did not constitute any remarkable place, but as population grew towards the end of the last century land revenue showed an upward trend, as is evident from Table 3.5.

Table 3.5: Sources of Revenue (in Rs.)

Year	Land revenue	Family taxes	Forest tolls
1874-75	38,789	34,931	28,395
1875-76	34,931	24,424	33,628
1876-77	31,421	25,415	22,955
1877-78	32,195	20,541	38,104
1878-79	33,963	19,705	39,572
1879-80	35,153	22,827	35,905
1880-81	34,717	18,630	37,944
1881-82	34,717	17,987	43,130
1882-83	29,394	24,224	48,594
1883-84	33,738	23,088	48,897
1884-85	35,295	23,587	56,902
1885-86	41,044	21,920	57,511
1886-87	40,656	23,840	89,520
1887-88	38,266	28,167	80,712
1888-89	52,018	20,986	76,796
1889-90	53,973	28,436	71,449
1903-04	2,32,000	NA	NA
1910-11	3,20,707	NA	NA
1911-12	3,31,974	NA	NA
1912-13	3,42,188	NA	NA

Sources: Compiled from—

(1) Administrative report of the political Agent, Hill Tipperah (1872-90) in DKC ed. AARPA, TSCRIM, Govt. of Tripura Vol. I & II.
(2) W.W. Hunter: Statistical Account of Bengal, Vol. VI (1973) p. 503.
(3) Education Directorate: Tripura State Gazette (compiled) ed. p. 43.

The table reveals that during the end of the previous century, there were much cultivable plainland uninhabited and shifting cultivation was preponderant among the hill tribes as the family taxes kept on rising, which resulted in a decrease in land revenue initially. Another reason for this decrease is the remission made by the king on account of the damage sustained by the crops due to natural calamities (viz., floods, famines etc.). However, the most glaring reason for this decline is perhaps the peculiar land system of the state. The farming out of land revenue from cultivable land to Thakurs/chiefs in return of nominal rent was the practice of the state for long; these holders usually 'settled' land with ryots who could not acquire occupancy rights. Consequently, quite a sumptuous amount of land revenue got lost in the midway. Even the Political Agent mentioned in his report that Behari coolies could be directly entrusted with the task of cultivating the not-so-fertile land by ploughing.[35] As revenue returns decreased considerably, land, related laws were passed in the 1880s—Rajasya Sambandhiya Niyamabali (1880); Praja—Bhumyadhikari Ain (1886) etc.[36] which changed the trend and land revenue gradually formed the greatest source of wealth since 1884-85 the land revenue increased, which again confirms the fact that immigration went on unchecked. Low revenue collection was checked with the giving away of taluki miyads; a considerable portion of land was brought under cultivation by ryots who received occupancy rights now. Yet the huge increase in land revenue during the early years of the present century coincides with the mass influx of plainlanders from across the border who were well versant with the plough, although the rate of rent was deliberately kept low by the government (Rs. 0-3-4 per high) in the remote interiors (0.62 per acre) to Rs. 1-7-4 per bigha for best quality land (Rs. 4.37 per acre).[37]

The fact that most of the immigrants were rural settlers dependent upon land can be seen from Table 3.6 which reveals the rural and urban population during the late 19th and early 20th century.

Table 3.6: Rural and Urban Population Distribution in Tripura (1881-1921)

Year	*Rural population*	*Urban population*
1881	95,637	N.A.
1891	1,37,442	N.A.
1901	1,63,812	9,513
1911	1,22,782	6,831
1921	2,96,694	7,743

Source: Extracts from 1931 Census reports of Bengal and Sikkim (Part I-Vol. V) by A.E. Porter, Published as Appendix IV to census of India 1961, Vol-XXVI, Tripura Part 1 (c) 1967, p. 522.

The table shows the swelling of population by 1,59,252 or 115 per cent during these years which cannot be by merely natural means and therefore, once again it can be said that the reason obviously lies in mass scale immigration only. The urban population during 1901-1911 shows a downward trend confirming the fact that more and more people tried to draw sustenance from land. The positive impact of growing cropped area, high land revenue collection soon gave way to negative impact. This jeopardised the interest of the local inhabitants of the state, shattered the remnants of isolation that they enjoyed as marshes, swamps got converted into firm and dry grounds which could not be used for shifting anymore, they started moving towards the interiors. The tribal population too increased with time, and land availability shrunk, land could not be left fallow for long, hence it could not regain its nutrient contents, consequently the slopes got eroded and forests too receded. The government did try some remedies–11,000 drones (28,490 ha.) in Khowai sub-division was set apart in 1930-31 (the Kalyanpur reserve)[38] similarly land was reclaimed in Kulai, Haor, Dhalai valley and 400 sq. miles of cultivable land was created to be distributed amongst peasants. But these lands were fit for ploughing and shifting was strictly prohibited. Thus the remedial measures of the government failed to encompass both the segments of the population which created a schism between the two communities, leading to grave implications in the future revolving around land question.

Socio-economic impact of the population pressure on land

Migration is considered to be critical because it changes the demographic profile as well as economic balance of the state as well as poses a threat to the socio-cultural identity of the region. The influx of neighbouring plainlanders into Tripura was on a much higher scale in comparison to the other states of the North-east, and it changed the state's set up in a very noticeable way, so much so that it seemed that the entire development of the state was the outcome of the settlement of the refugees in large numbers.[39] Absence of industries or any subsidiary occupation force the people to fall back upon agriculture as their only source of livelihood. With the rising pressure of population, the density of population too rises which implies a low man-land ratio as a result of systematic sub-division of holdings. The density of population per square kilometre has been on the rise since the last quarter of the century, in 1872 the ratio was 3.36 per square km. which rose to 16.54 at the turn of the century to 109 during 1951-61 and reached a high of 263 according to the 1991 census.[40] This increase was reflected directly in the rising trend of uneconomic, marginal holdings i.e. holdings with an area of 1.0 hectare or less as also a fluctuating trend in their total operated area. Available statistical data from the Agricultural Census[41] conducted in 1970-71 shows a huge number of such holding, 1,73,790 or 69.5 per cent of the total holdings which operated on about 27.6 per cent of cultivable land or 70,170 hectares. This shows that regular transfers, forced sales, partition, formation of new families increased the number of marginal farms which points out to the obvious population pressure on land arising out of massive influx. By this time urbanisation had gained momentum and cultivable land near the towns were reclaimed for development purpose which affected the holdings and fragmented the plots, 1976-77 Agricultural Census shows a decline in number of marginal holdings–1,44,897 (68.8 per cent) and even the operated area decreased–66,449 hectares (21.5 per cent of cultivable land) it is due to the fact that holdings with negligible area of kitchen gardens by non-agriculturists was excluded. But the next span (1980-81) witnessed an increase in both number of marginal holdings as well as total operated area i.e. 2,10,984 (68.6 per cent) holdings and 1,01,983 hectares (30.9 per cent), this is perhaps due to the imposition of ceiling, acquisition of khas land by encroachment or simply possessing them. After that the numbers have been constantly increasing as is reflected in the data of 1985-86 (2,11,302) and in 1990-91 (2,16,826). Presumably, such small farms reduce the disparities that exist in the distribution of land in the state and also make land available for the rehabilitation of landless agriculturists. But in a state where cultivable land is only 11,000 ha. of the total available land, such low man-land ratio (.38 ha. at present) naturally affects the economy adversely.

Not all who received land or got to 'own' them could enjoy their rights and some had to relinquish their hold on the same and turn agricultural labourers. Within a few years there was a considerable growth in the number of agricultural labourers in comparison to agriculturists. The labourers constituted 8.93 per cent (1941-51) which decreased to 7.53 per cent (1951-61) but then it rose to a high of 19.26 percent (1961-71) to 23.91 per cent (1971-81), and 23.38 per cent in 1981-91 while cultivators declined inversely, 62.94 per cent (1941-51), 64.24 per cent (1951-61), 54.41 per cent (1961-71) to 43.57 per cent (1971-81) and 38.09 per cent (1981-91). Interestingly the decade following partition saw a decrease in labourers and an increase in cultivators. Analysing the various reasons it can be said that as the influx gathered momentum, land was reclaimed or 'created' to settle these 'new settlers' and thus the number of cultivators increased. Another strange development accounts for the increase of agricultural labourers among the autochthons which was unheard of so long. As the displaced people got 'settled; the indigeneous people with land (either lungas or plains) were compelled to lease them to the new settlers who could use them better. But such land being mostly a part of community land with no permanent, heritable right of the owner, the leaseholders gradually acquired the land taking advantages of the anomalies of ownership rights. And the simple minded original land owners either had to retreat more into the interiors to pursue uneconomic shifting or crowd the already overcrowded land market as agricultural labourers for survival. The schism, which was created with mass immigration further deepened bringing along more complications.

Population pressure on land has also brought considerable unemployment in the rural sector. The nature of unemployment here may be categorised into two sects–seasonal and chronic. Seasonal unemployment is the result of shifting cultivation and the method of single cropping pattern followed thereof. Chronic unemployment comes in the form of disguised unemployment which may not be apparent always. The fast increase of essentially land dependent population without a corresponding increase in availability of land or landrelated jobs obviously results in a surplus population engaged in agriculture. This becomes more obvious in case of family ownership, where a plot of land is tilled by the entire family, a job which any one or two member could do satisfactorily, as the yield remains the same. It implies that the workers are working below their capacity, though apparently all of them are employed. This is underemployment of a perennial nature, which portrays the dismal picture of the economy wherein a large number of able-bodied workers cannot find any suitable job throughout the year and inevitably falls into the clutches of moneylenders for survival.

In an agriculture dependent state, it is but natural that the environment needs to be maintained and food production enhanced to cater to the needs of the people. But due to the aforesaid reasons it was found that those self-sufficient shifting cultivators could not produce enough to last them for a year, as shifting became uneconomic. The craze for plough cultivation lured the better–off shifting cultivators who left their age old occupation for better prospects and shifted to the plains. However, ignorance and lack of skill hindered them from producing enough to suffice their needs even. Meanwhile shifting did not stop altogether but as land was being reclaimed (swamps, marshes etc.) it became unviable and soil erosion continued unabated destroying the nutrient contents of the soil. Grazing too led to colossal loss of land in respect to its fertility and productive capacity. All these accounted to such a huge proportion that almost 70 per cent of available cultivable land has been affected and the rest is deteriorating fast too. Forests are felled indiscriminately, trees cut down which recedes the greenery at a direct proportion with development and this has a severe impact upon the ecological balance of the state, even the mode of food production needs to undergo a change. The border area are worst affected as quite often people from the other side cross over unrestricted in search of greener pastures and soon arable land turn into barren tracts.

Not that the government has turned a blind eye to these grave problems all these years, but the huge army of unemployed and underemployed people are continuously growing with the rapid population growth. The local inhabitants enjoyed communal ownership of land and in the absence of any land records they rely upon customs handed down from generations. Though population explosion has been considerably checked, the two communities cannot be separated as after living alongside so long both can be termed as 'sons of soil'. The tribal customs have to be maintained as well as government's policy of setting the plainlanders have to be adhered to. Right of ownership encourages the owner to adopt measures for permanent improvement of land, any slackness caused by insecurity, fear of losing the land results in low productivity. Concentration of money in a few hands encourages absentee landlordism, an emerging feature at present and their ill-effects are well known as they do not improve agriculture. The local inhabitants dispossessed of what they considered to be their own, the feeling of eviction gave vent to their pent-up hatred, fanned by political parties and communal tiffs became a common occurrence ultimately leading to a number of people turning militants.

The population pressure on land is acting as an impediment to the economic development of the state. Agriculture is the predominant means of livelihood for thousands of people of this state, be it shifting or ploughing type of cultivation. Here a generalisation between the two types of cultivation is simply impossible and so a balance between the two is to be maintained. The autochthons were quite content with their age-old mode of shifting cultivation, but as the population increased, their productivity decreased and they had to move further beyond losing whatever they had as their own while witnessing the emergence of private property amongst the settlers as well as better-off autochthons. The cause of tribal unrest can be attributed to mainly population increase and their tendency to colonise (settle). Ester Boserup[44] points out that with increasing population, there has been a shift in recent decades from more extensive to intensive systems of land use, population pressure raises the efficiency of labour and opens up opportunities for a more intricate division of labour. Such hypotheses can be applied to Tripura too, but owing to its undulating topography, bottlenecks in infrastructure, insecure tenancy systems, lack of credit facilities etc. the agricultural sector remains underdeveloped. Out of a total area of 10,490 sq. km, 606 sq. km, is more or less cultivable, that too out of this area 7 per cent is irrigable. Despite such handicaps, Tripura's huge population of cultivators, perhaps could have helped in development, but their ignorance acted as an impediment too. The increasing population pressure destroyed the traditional peasant economy, its associated communal structures, rights, customs and practices. Class differentiation, a totally new phenomenon has emerged among the local people, spoiling the remnants of collective element in their society. High density of population, leading to low man-land ratio, relentless destruction of flora and fauna depletes the non-renewable natural resources continuously affecting the bio-geo-chemical cycles of the nature. To get maximum output from a limited resource input, the entire agricultural framework needs to be recast, the land reform policy in the hills has to be undertaken in a wider scale, and not limited to the general idea of fixing land ceilings and redistributing land. An overhauling of the land reform which ensures maximum investment, makes full land utilisation possible irrespective of who the man is behind the plough can impart dynamism in this essentially peasant economy.

NOTES AND REFERENCES

1. Census of India 1991 Series 1 Para 2 of 1992 (Final Population Totals) Statistics Department, Govt. of Tripura.
2. Barkataki S. (ed.): *Tribes of Assam*, New Delhi, National Book Trust, 1962, pp. 1-2.
3. Hunter, W.W.: *A Statistical Account of Bengal* Vol. VI, London (1870) (reprinted 1973) p. 492.
4. Choudhury D.K. (ed.): *Annual Administrative Report of the Political Agency*, 1872-74 & 1889-90, Vol. I, Tribal Research Centre, Agartala, (1994) pp. IV-V. The first recorded estimate of the population of Tripura is available for 1872, which was considered as an underestimate in the later years since the system of data collection regarding population was still unorganized, its importance not understood and hence estimate made in a casual manner.
5. Hunter W.W.: *op.cit.*, p. 502.
6. Singh, K.: *India and the South Asia*, a Seminar Paper presented at the University of Rajasthan, Jaipur on Trans-Border Migration, Mar., 23-25, 1987.
7. *Annual Administrative Reports of the Political Agency* (1872): An unpublished report on the influx of the Mogs found at the West Bengal State Archives.

8. *Tripura Administrative Reports* (1343 T.E.): Unpublished Administrative reports of then State Govt. found at the Tripura State Archives.

9. *Journal of Asiatic Society of Bengal*, Vol. XIX, 1850, p. 556, (We learn this from a booklet published on the occassion of an exhibition of Indian Coins. Organised by the Calcutta Coin Society at the Asiatic Society, Calcutta in 1974).

10. Datta, B.C.: *Thirty years in Tripura* (Dharmanagar division) in Bengali, Agartala p. 11.

11. Deb Barma S.C.: *Census Repot* 1340 T.E. (Compiled) Agartala, 1933, pp. 51-57.

12. *Ibid.*, p. 45.

13. The Imperial Gazetteer of India, Vol. XIII (OUP), London 1908, p. 119.

14. Hunter: *op.cit.*, p. 513 (The Praja Bhumyadhikari Ain 1886).

15. Gon Choudhury (ed.), *Tripura: The Land and Its People,* Leela Devi Publications, Delhi, 1980. Apaper by N.C. Dutta on Land Reforms, pp. 105-123.

16. Dutta, D.C. and Banerjee, S. (ed.): Rajgi Tripurar Sarkari Bangla, Education Directorate, Govt. of Tripura, 1976 (A few unpublished documents by late Prof. D.C. Bhattacherjee) pp. 500-540.

17. Ganguly, J.B.: *The Benign Hills,* Tripura Darpan Prakashani, Agartala (1983), p. 21.

18. Chandra Bipan: *Modern India,* NCERT, New Delhi (1976) p. 185.

19. *Ibid.*

20. Dutta D.C. & Banerjee, S. (ed.): *op.cit.,* p. 593.

21. Singha K.C. Rajmala (4th Lahar): Published by the Education Directorate, Tripura, 1976, pp. 134-135. (Kalyan Manikya granted one Vishwanath Sarma, a priest, 5 drone land to defray the expanses of a temple of Bhagavati at Kailagad).

22. Dutta & Banerjee (ed): *op.cit.,* p. 593.

23. *Ibid.*

24. D. Choudhuri, D.K. (ed.): AARPA, *op.cit.,* p. vi.

25. Dutta & Banerjee (ed): *op.cit.* p. 953 (Asma Sanad-land given to Muslims for religious purposes).

26. *Ibid.*

27. Talukdar, S.P.: *The Chakma Life and Struggle,* Gyan Publishing House, Delhi, 1988, pp. 47-58.

28. Gokhale, N.A.: *Chakmas—A persecuted Lot,* The Sentinel, 17 Dec., 1986, p. 3.

29. Informations as gathered from the Directorate of Refugee Rehabilitation, Government of Tripura (1993 Nov.).
30. *The Stateman News Service: The Statesman* (Cal.) Apr. 12, 1993.
31. F.No. 2509-10/DM/LA/SS/60/64, Tripura State Archives (unpublished).
32. D.O. No. 10471/DM/LA/57 (unpublished).
33. Excerpt from the letter of Assistant District Magistrate (Unpublished), TSA.
34. Report of N.C.A.E.R., New Delhi, 1961, p. 8.
35. Choudhury, D.K.: *op.cit.,* p. 28.
36. These laws dealt with land tax from Kayemi Taluk (i.e. permanently settled Estates): Khas mahal (i.e. Government owned land), Karsha praja ((i.e. cultivating tenants) and the relationship between Talukdars (i.e. Landlords) and the Raiyots (i.e. Tenants).
37. Report of the Land Revenue Commission, Bengal, Vol. II Table-VI (a) Government of Bengal, Appendix in, pp. 100-104.
38. Menon, K.D.: *Tripura District Gazetteer,* Education Directorate, 1974, pp. 60-61.
39. Burman, B.K.: Demographic and Socio-Economic Profiles of the Hill Area of the North-east India, Census of India, 1961, New Delhi (1970) p. 104.
40. Imperial Gazetteer of India, Vol. XIII, 1908, p. 506 (Extracts published in Census of India, 1961, Vol. XXVI Tripura Part 1 (c) 1967).
41. State Report on Agricultural Census in Tripura, Agricultural Census Commissioner (Directorate of Land Records and Settlement), Agartala, 1980-81.
42. All India Bulletin on Agricultural Census, Department of Agriculture and Co-operation, 1985-86 and the basic Statistics of North-east, North Eastern Council, Shillong, 1995.
43. Census extracts, various issues.
44. Boserup: E. *The Conditions of Agricultural Growth,* Aldine Publishing Co., Chicago (1965), pp. 200-240.

4

Shifting to Plough Cultivation: A Transition in the Economy

Analysing the history of development it is found that man started as a hunter and food gatherer and then turned to domestication of animals for drawing sustenance. With growing population pressure, these means of livelihood seemed inadequate and thus he embarked upon a rudimentary form of agriculture, widely known as "shifting cultivation". The historical origins of this mode of cultivation can be traced back to the Neolithic period (13,000 B.C. - 3000 B.C.),[1] when man made a group comprising of his clansmen for better social and economic security and resorted to agricultural activities though he retained his nomadic habits too. This form of cultivation has been defined as follows: "Shifting cultivation is the custom of cultivating clearings scattered in the reservoir of natural vegetation (forest or grassland) and of abandoning them as soon as soil is exhausted; and this includes the custom of shifting homesteads in order to follow the cultivator's search for new fertile land"[2]. Geographically, this practice of shifting cultivation spreads throughout the East-West stretch of the globe within the tropics and sub-tropical expanse,[3] where heavy rainfall and heat usually favours and nourishes such agriculture (countries like West Africa, France, Belgium, Ceylon, Japan, Indonesia, Phillippines). In India this mode is rampant in Madhya Pradesh, Bihar, Orissa, Andhra Pradesh, parts of Kerala, Karnataka and in the North-Eastern states where it has acquired gigantic proportions. Approximately a minimum area of 14,660 sq.km or 5.7 per cent[4] of the total land available is under shifting cultivation at one time or another in this region, though the people are exposed to modern techniques since the beginning of the present century.

In the past, this mode was quite viable as there were hot and humid forests with luxurient growth of vegetation, low pressure of population even though a majority depended upon shifting. However, shifting cannot produce in large scales and so there always existed a development potential concealed within this sector, in the form of surplus labour and unused productive capacity of tribal land.[5] The dependence on this primitive mode of cultivation is primarily because of the hilly topography and the existing socio-economic conditions of the inhabitants. An estimate of the extent of shifting cultivation in this region can explain the dependence.

Table 4.1 reveals that at any given period of time (1983 in this case) the annual area under shifting cultivation takes up 1.5 per cent of the total geographical area of the region (3869 sq.kms). While Manipur records the largest area (900 sq.km), Nagaland maintains the smallest area (190 sq.km) under shifting. The fallow period required for the land to regain its nutrients also vary but a quick glance gives the idea that the fallow period has decreased over the years. Even the minimum area under shifting is highest in Manipur (3600 sq.km) while it is the lowest in Tripura (1115 sq. km) followed by Assam (1392 sq.km) perhaps owing to the spread of settled cultivation as also presence of a large number of plainsmen who are not particularly adept at shifting. In Nagaland quite a huge number of families are dependent upon shifting (1,16,046) though the area affected is the least (190 sq. km); while only 7000 families of Manipur carry on shifting cultivation where the area affected is also high confirming the fact that shifting cultivation is widespread. Taking the North-eastern region as a whole we find that 5.7 per cent of the total geographical area is under shifting cultivation at a time (14,660 sq. km) involving a total of 4,43,336 families, which is not at all a small figure. Repeated shifting for long has devastating effects on land which leads to heavy soil erosion, a total of 702 sq. km of forest cover is lost due to shifting (1993 Basic Statistics of North-east Region) and is responsible for floods in the lower reaches of the main river systems;[6] as also to loss of fertility and productivity of the soil. Yet shifting cannot be stopped here as it is regarded as a reflex to the physiographical character of land, and a sentimental value is attached to land as in Meghalaya and Mizoram. Though plough has been introduced in Nagaland and found quite yielding too; terrace cultivation acts as a substitute in Arunachal Pradesh shifting is not altogether discarded.

Table 4.1: Shifting Cultivation in the North-East Region

States	*Annual area under shifting cultivation (sq.km)*	*Fallow period (Years)*	*Minimum area under shifting at a time (sq.km)*	*Number of families involved*
1. Arunachal Pradesh	–700	3-10	2,100	54,000
2. Assam	–696	2-10	1,392	58,000
3. Manipur	–900	4-7	3,600	7,000
4. Meghalaya	–530	5-7	2,650	52,290
5. Mizoram	–630	3-4	1,890	50,000
6. Nagaland	–190	5-8	1,913	1,16,046
7. Tripura	–223	5-9	1,115	43,000
Total	3,869 (1.5 per cent)		14,660 (5.7 per cent)	4,43,336

Source : Task Force on Shifting Cultivation Ministry of Agriculture (1983) Quoted from Basic Statistics of the North-East Region (1995) North Eastern Council, Shillong, p. 30.

However, in the case of Assam and Tripura, the case is somewhat different due to its demographic character resulting from mass influx of immigrants from the other side of the border. In Tripura's case this is more glaring because, being the tiniest of the states, it has valley land too, immigrants are plainlanders and so shifting cultivation could have been arrested to quite an extent. Immigrations initially brought down the barriers between the plainlander's advanced techniques of cultivation and indigenous tribal's subsistence mode of cultivation yet a huge percentage of cultivators are still engaged in shifting mode. An entirely new stage of development, which can be termed as a 'transitionary phase, an unique feature of agricultural dualism of 'traditionality versus modernity[7] came to the forefront. This chapter attempts at bringing out this transition after the coming of the plainlanders.

Shifting cultivation as seen is an agricultural system in which plots of land are cleared by setting fire and are cropped discontinuously, keeping land fallow for longer periods than they are cropped. The early inhabitants of Tripura came in search of land to pursue age-old shifting cultivation. They were the Tipras, Jamatiyas, Chakmas, Reangs, Noatias etc. Even Webster supports this fact "the Tipras first settled in the plains of Sylhet and in the northern portion of the hills of Tripura 1500 yeas ago.[8] That they pursued shifting can be seen from the fact that though there were long strips of fertile lands abundant in the districts of Sylhet, Tipperah and Noakhali where plough could be extended such cultivation cannot be found.[9] Therefore, the simple autochthons depended upon the primitive mode of cultivation since long. The implements too remained simple all these years, a 'takhal' is used which suffices to be both a hoe and cutter; 'Cekhras' are used which are sickles used during harvesting time and baskets serve to carry the seeds and the products. The hot, humid forests with luxurient growth of vegetation and largely scattered small, human habitation are the ideal setting for the same. The process is simple too and may be divided into three stages. The first is selection of a virgin slope with bushes and plants and then leaving it to dry for some days. Certain sacred rites and rituals are performed by the village elders, sacrifices offered to their deities, super-natural spirits. The slope is then set on fire which is kept burning for a few days. This ash is considered to be a good manure and the entire slope is covered by the same. This phase takes place during spring i.e. March to April. The second state is the actual cultivation phase which starts during summer as a few showers are needed during the sowing of the seeds (June-July). The huts are built on an elevated ground so that the cultivators can keep vigil on their crops to protect it from wild beasts, weeding is done during this time. The third is the harvesting stage (September-October), which is an elaborate ritual too including worship of Bura Devta to whom sacrifice is to be made, and only after that do they consume their products. The main products are paddy, a variety of vegetables, sesamum, jute and cotton.

The kings of Tripura depended upon the rich plains of their zamindari and made the hills of the kingdom subsequent to and dependent upon the British economy. Those inhabiting the hills were nomadic in nature and knew only shifting mode as a food producing method and since these areas were inaccessible not yielding much revenue they were looked upon as something apart. The indigenous population comprised of 36 per cent of the total population during 1874-75 who were subjected to both immigration and emigration.[10] At that time shifting seemed rather viable and appropriate too since ploughing was not possible on the slopes where the top soil would be washed away along with the rains. Gradually this got so much embedded into the lives of the people that their daily activities, beliefs, attitudes, values revolved around it. Ploughing, though not totally unknown, did not spread fast as the *Imperial Gazette* reported as late as 1908 that the "nomadic tillage, known as jhum, had acquired enormous proportions in the hills of Tripura and formed an important part of the indigeneous peoples' traditions, perceptions and socio-economic relations", or, to put it in another way, jhum was the predominant cultivation and it occupied a vital position in the state economy. It is also seen from the administrative reports of the last quarter of 19th century that taxes levied on shifting, family tax, or on its products i.e. cotton tax constituted the chief items of revenue. It is also reported that even if plough is introduced it cannot be much successful as it had not worked in the Chittagong Hill Tracts too.

With abundant land, the hill dwellers had no problems in pursuing shifting mode of cultivation widely. There were no strict boundary demarcation and hence the various tribes could easily 'shift' continuously to hills and slopes as they fancied. Crops were able to use the nutrients sufficiently even solar energy could be used better and land saved from erosion in the undulating terrains. Moreover, different crops have different growth habits and requirements, those harvested leave behind plant - debris which decay and provide organic matter to the standing crops, and thus optimally utilises the micro-climatic and seasonal changes with the cropping variety. There were no regular system of land tenure in the hills, no rent was demanded by the ruling elite either except for certain taxes on products and the family tax; the cultivation mode required no hired labour, fertilisers or expensive implements. Therefore, initially the yields were quite high considering the fact that cost of production was low and there was ample land fit for shifting even within the limits of the state.[12] This cultivation got itself harmoniously blended into the traditional society and besides providing an economic base for the community acted as a strong binding force too.

The New Contact

This harmony could not continue for long. The British authorities pressurized the kings to pay more revenue, the latter began to search for new avenues where in plough cultivation was encouraged, small plots (as taluks) were leased out and the shifting cultivators were left outside the ambit of development. Plough cultivation had to reclaim land even in the foothills of the hillocks, shifting began to be discarded here, moreover, the shifting cultivators were increasingly being taxed too. This threatened the basic aspects of tribal polity as those so long engaged in shifting mode were gradually taking on settled cultivation while retaining shifting only as a subsidiary occupation (e.g. Mogs, Chakmas).[13] The Manipuri community, though not in favour of shifting cultivation, took up land under Jangal Abadi system for plough cultivation where in no rent was charged for 4-10 years and even after that the rents were kept quite low in the plains. Forest resources were now considered to be valuable assets and shifting came to be prohibited here and those involved were turned out of the forest vicinities so that they may not harm the ecological balance. Later the Jiratia tenants flocked in (details in ch. 4) who cultivated the plainsland in the border areas. Land for shifting cultivation was perpetually reducing and considering their production potentials during this time, the burden of taxation was much higher upon the hillsmen, they had to retreat towards the interiors as land seemed to fizzle out of their hands, and comparatively limited area was subjected to such cultivation tactics.

Dwindling of land for shifting made the cultivation unviable as land carrying capacity declined, their shifting became limited and fallow periods got drastically reduced. In regions where shifting was done on steep slopes, soil became porous and could not retain rain water which floated down with loosened top soil. Hence their nomadic habits had to be given up, which was a land-mark change in their lifestyle. Even the rulers of the state tried to alleviate the poverty of these cultivators by adapting them to either plough or some other subsidiary occupation. To assist the shifting cultivators to switch over to a life of comfort certain areas in the state were set apart during 1930-31.[14] About 11,000 drones (70,400 acres) in Khowai sub-division was reserved for the shifting cultivators to pursue plough and this was later increased to 12,48,000 acres in 1941. But the progress was not found to be satisfactory as is depicted here - "Many families have since permanently settled in the reserve once and is taking to ploughing, though the area actually brought under cultivation during the period is not very encouraging. In the year 1347 T.E. 16 drones and 15 kanis[15] and in 1348 T.E., 10 drones 3 kanis of land came under actual settlement in the area." The Census of 1931 shows that out of 69,868 main workers who adopted agriculture and related occupation, 56.79 per cent were settled owner-cultivators while 27.01 per cent were still doing jhum, and 55.52 per cent tookup shifting as a subsidiary occupation.[16] But they did not take the pains to inculcate this idea in the simple minds of the tribals which left them stranded in mid-way. They started producing cash crops like jute and cotton[17] while rice was procured from plough, but the bulk of shifting cultivators did not have the means to follow such specialisation and remained poverty-stricken. Since then shifting has changed to the status of subsidiary occupation and the rites and rituals associated with the same have also gradually disappeared as a natural sequel to it.

By this time, these traditional custom-bound tribesmen were exposed to the relatively advanced economy of the plainlanders which ushered in a new phase in the economy of the land. The self-contained, subsistent economy was pitted against the economic set up of the non-tribal settled people which was hierarchical in structure. Thus, this transition from a self-sufficient, simple, non-hierarchical barter economy to a more broader, heterogenous, stratified, money economy remains quite a complex process.

It is well known that shifting cultivation acts as a strong binding force for community life, where in natural resources provide a common source of livelihood to which all the members of the community have an equal right. This concept of community rights weakened gradually on two counts. The advent of plainlanders brought in certain new rules of land, emergence of a feudal gentry and their claim on individual ownership rights. The social organisation of the people was also built around the same concepts of community participation. But the exploiters who came in the guise of money lenders, land-hungry peasants, petty traders resulted in the break-down of the tribal land-system and the first impact could be noticed in the formal establishment of private property in land, commercialisation of agriculture on one hand and the governmental protection on the other shaped the process of transition. They were becoming well conversant with money economy as they realised the worth of coins as a medium of exchange.[18] A large section of well-to-do shifting cultivators, like the chiefs and their kinsmen opted for settled cultivation as they understood that to remain on the same plane as their non-tribal counterparts they had to 'earn' too. Moreover, the rural development programmes of the government also helped the individual in creating individual assets. At the same time, there is another section of these shifting cultivators who felt shy of the change and could not accustom themselves to this new socio-economic set-up and so retroceded their steps towards the remote hills. Perhaps they were afraid of being completely cut off from the traditional society. In so doing, they became estranged from their mainstream life and were left out in the process of development too.

The Effect of the Transition

The increasing contact with the outside world and the adaptation of the shifting cultivators to a new life style has brought with it a number of hazards. There has emerged diverse economic needs in the traditional societies. The hitherto self-contained economy was able to cater to most of their needs but now they became conscious of what money could buy an entry to better living. They tried to change over to settled cultivations overcoming the psychological barriers. But already land available for the same had become scarce, they were only in the form of strips along rivulets or the swampy lunga land, and here too success was not easy as they were not quite familiar with the method, the implements were inaccessible and so crops failed often. And as the community factor was already obsolete they did not have anything to fall back upon and resorted to selling off the meagre holdings even, or try their luck at some other menial occupation. The cohesive tie between the kinsmen, the sense of mutual cooperation were lost in the process of transition.

In this new contact process, land for permanent cultivation became a key factor.[19] Those with land acquired greater economic power which led to the emergence of a tribal aristocracy hitherto unknown. The kins of the former chiefs who enjoyed special status during princely rule, could gain power and wield authority during this process. The permanent fields needed more labour which family labour could not suffice and hence labour could easily be roped in from his own community or outside too. This new form of relationship within one's own community threatened the egalitarian socio-economic structure of the traditional society. Gradually a difference in their status came to be noticed, different from those who still had to pursue unviable shifting unable to acquire enough asset and procure the most important resource, i.e. land. Thus, two distinct sections are discernible among the shifting cultivators during this transitionary process - those who are completely dependent on shifting and those who own permanent fields but had not forsaken their shifting rights too. As the shifting cycles decreased, the former group faced dire poverty as production fell and land grew scarce too. They are the weaker group undoubtedly and do not have access to any beneficiary programmes even which is taken up by the latter group, consolidating their individual ownership claims on the lands over which they enjoyed only usufructuary rights.

The labour involved in shifting cultivation is another fact of the transition which needs attention too. Labour in the traditional shifting society is based on mutual help and cooperation, whenever any of the members need help in preparing the field, building of a hut or any similar need he would easily seek the help of other fellow members as all have equal access to the natural resources, and no differentiation was made. As the pressure of population grew, there began a hunt for alternative means and members of the community strove hard to get hold of any available means of sustenance. The labour which they readily gave in the past for no minimal returns was now being treated as a productive asset. At this stage, mutual help and cooperation broke down and a new form of labour relationship resulted which shattered the remnants of traditional society.

The sociological parameters of the people's life is inter-woven with their form of cultivation. The social system is based on equality wherein all the members could enjoy full social security built in their economic system. Their social relationships are embedded in primitive kinship bonds which mould them culturally. It is found that though the cultivators are quire aware of the unviability of their production method and the economic benefits of settled cultivation, the social conditioning of their life poses a constraint in the way of adaptation. Their age-old customs, rites and rituals their sentimental attachment to the land they cultivate, the element of cooperation involved all act as hindrances. The change in their economic system has made their social system respond in its own, but the change is quite sudden and drastic with serious sociological consequences, which leaves the option open for the community to either accept or reject it. In this milieu, there are always some who may be guided by personal interests in taking advantage of the new opportunity. This is a limited group who ultimately take over the leadership while the others who are not fortunate enough fall prey to the constraints of change.

This transition has social ramifications too. A very important change has occurred in the ethnic composition of the population with the opening up of the traditional economy. A number of tribes assimilate in their pursuit of adopting settled life, also there are non-tribal settlements in the fringes of these areas. Consequently, there is a tendency towards social and cultural integrations with the non-indigeneous inhabitants of the soil. But in reality, this integration is often short-lived and the reciprocal cordial relationship once shared by the communities cooled down in a short time. The people have become self-sufficient in a sense that they do not have to depend upon their kith and kin to the same extent as before. The new farmers are able to convert their surplus products into cash, a direct effect of commercialisation. Formerly food surplus were used to increase social interdependence, cultivating the feeling of good neighbourliness, but in a changed atmosphere such feelings became a thing of the past and consequently, they are in an uncertain phase neither are they welcome in the alien company, nor can they retrocede, New laws, conventions, litigations regarding land baffle them, which even lead to fragmentation of land a fact which they never heard of. Those who were eager to initiate the new culture and socio-economic set up tried to identify themselves with the dominant group. They came out of their cocoons into the big world but found themselves being discriminated against. They have to suffer from under-employment or even unemployment in the new surroundings, and often revert back to shifting which seemed to be no cumbersome process in regard to labour or capital.

Generally, the shifting cultivators are often in need of credit, introduction of money economy led them to a spending spree when they received money as rent or sold their land. The festivities, which were simple affairs till now began to be celebrated with great pomp and show, and so their borrowings too increased. Moreover, as shifting cultivators they could not store any surplus produce for want of accommodation and so occasionally had feasts too, as settled cultivators they could not forgo the same and spent lavishly. This credit was usually supplied by the money-lenders in the form of a crop advance called dadan. This is a tempting method of purchasing the crops of commercial importance from the jhumias on very advantageous terms[20] and the repayment was in kind of fixed rate. Needless to mention, it usually went against the interest of the cultivators and they had to pay a very high rate of interest too. Even during transition the position changed little, the condition of repayment is an agreed sum paid in a jhum crop calculated at the going market price of that crop. The cultivators always borrow from the village money-lender and the process is a recurring feature in their lives. Indebtedness was not heard of in their society but now whenever the need for money arises they do not hesitate to borrow from the local money-lender. And the repayment may be anything, cash or kind, and for that they often have to sell off their agricultural or even homestead land and be rendered landless. The benchmark survey (1978)[21] shows that during a period of four decades 1938-78 almost 3000 households transferred their land to non-tribals because of failure to repay the loans. Almost 7 per cent of the jhumia villages had to give up settled cultivation and go back to jhuming[22]. Such evidences bring forth the fact that transition has been a rude shock to the simple jhumias though initially they were attracted to the same.

As the plan period was initiated (1951) special attention was given to the shifting cultivators, where in the commission decided to introduce improved method of agriculture as well as rehabilitate them.[23] Tripura government started a pilot scheme in the Belonia sub-division (1953) and "it was decided that every jhumia family should be given land and induced to lead settled life by adopting cultivation with the plough"[24] and from then on such schemes are going on in full swing. The first was set up at Bishramganj in January 1957 and 9 colonies (848 families) followed till 1957-58. Land, money-grant, housing facilities were given, training in artisan, carpentry etc. provided, multipurpose societies setup to supply the essentials, educational opportunities, medical benefits given, seeds, fruit - graft supplied, fisheries, irrigation projects made available, even the Reserve forests were used to settle such cultivators.[25] Till 1958 13 forest villages were set up. But soon after, certain colonies began to be deserted as it was found out that it imposed certain constraints on their life style. They could not pursue shifting here and the jobs they had to indulge in only suited the ablebodied young men while the elderly had to remain content with the cash-grants or pursue settled cultivation which they could not properly utilise too. Moreover, the tribal families usually practice the norm of setting up separate household once the young gets married, the assimilation of two lifestyles left the older generation alone to fend for themselves as the former felt no obligation to look after the aged. Some even had to leave the colonies and once outside had to revert to shifting again which collapsed their social security system. A basic reason which caused rampant desertion is inadequate recognition given to the complexities arising in the transitionary process which brought a structural change in the economy as a whole. Though the basic elements involved in both the modes of cultivation remains the same there is a totally different method involved which has been overlooked by the administrators. The hill tribes were suddenly introduced to settled cultivation without being prepared for the same, the essentially jhumiya has never held the plough or even managed the bullocks before, and so cannot gather the skill of agricultural operations. And such inability of his has been ascribed to lack of interest, incapable of adaptation, and so on. It should be remembered that it takes quite sometime to change over from one occupation to another, to acclamatise oneself and gain experience. The autochthons like to be surrounded by the forests, their own totems and deities, but settlement schemes did not look into these aspects of their lifestyle. Consequently, the jhumiya viewed colony life as alien, and felt ill-at-ease in the new setting. They did not particularly feel at home in the houses built for them as they did not cater to their needs, and sometimes even built their familiar hutments in the vicinity of the colonies leaving the allotted houses unused. During their stay in the colonies, any mishap or accident was attributed to earning of the Almighty's wrath on account of their changed way of life.

The question regarding sustenance of the shifting cultivators during the transition needs to be mentioned here. The nominal cashgrant hardly provide for the implements required and bare essentials during the gestation period, so the early years are problem-ridden. Tremendous efforts is needed in establishing an effective hold on the land which initially does not match the labour input. Therefore, unless the government can provide for the necessities during this period the cultivation is likely to face the exigencies of a bad year (pestilence, crop failure, rain damage etc.) which either makes him abandon his work to seek others or reverts him back to age-old self-sufficient shifting. Another anomaly which came to the notice was that of official discrepancies. It was found that those with acquaintances or political links in the government could avail of the subsidies more often than others. They were reluctant to build up these comforts once again rather could forgo this sheltered life even. The colony life did not pay heed to the class consciousness of the tribes - all were huddled together irrespective of their class status. The chiefs felt disgusted at such treatment, even the inmates who so long depended upon the chief as their sole leader were now confused about whom to resort to in distress, the chief or the administration? It was soon found that if one disgruntled headman left the colony, his clansmen soon followed him out. Some complain about not being able to become the actual "owner" of the land as "pattas" were not given to them. Such acts of the government often lead to frustration among them. There are reports of tribals not being in round-the-year occupation and preferred the easy way of shifting. Despite the efforts made by the government, shifting cultivation could hardly be stopped or even the numbers decreased. The Tribal Welfare Department came up with some data regarding the increase of shifting cultivators and at the same time decreasing of land available for the same during the decade 1960-1970. Where as those involved rose from 28,554 to 35,000 the area where they carried on their job shrunk from 6,73,734 ha to 5,01,731 ha. This is a clear evidence of the unviability and uneconomic character of shifting cultivation, yet owing to the peculiar characteristics of the agrarian structure shifting could hardly be arrested.

The transition from shifting to settled life, therefore, has not been a smooth one as is evident from the aforesaid analysis and there have been repetition of the same mistakes in rehabilitation projects time and again. The shifting cultivators try to go back to their age-old occupation whenever in difficulty. To make amends settlement schemes have been diversified since the end of the Sixth

Plan[26] (a) For agriculture based scheme, 0.80 ha of land is allotted per beneficiary of which 0.40 ha is to be lunga land; (b) For horticulture purpose, 2 ha of land, (c) for pisciculture purpose, 630 ha; 50 per cent of the cost of land (not exceeding Rs. 5,900/-) is given as grant and the remaining 50 per cent as a loan by the Scheduled Tribes Corporation. The tribal rehabilitation in plantations and primitive group programmes have helped in diverting traditional jhum cultivation by introducing plantation crops and fruit species by utilising barren hillocks. Even the Tripura Rehabilitation and Plantation Corporation Ltd[27] introduced rubber plantations in compact blocks on land granted by the government on the basis of "owner-cum-worker" plans. These depict the fact that now weaning the jhumias to settled type of agriculture is always not feasible due to scarcity of resources. And at the same time the administration cannot wait for shifting cultivators to "open up" and aspire for change, which will go against planned economic development. With growing pressure of population, shifting cultivation has become ecologically destructive and economically unviable too, and the cultivator himself is aware of this fact and it is not that he does not want to change. However dearth of fertile land, over populated regions and lack of initiative on the part of the authority often bring back the shifting cultivators to their old-forsaken job. The administrative machinery of the state since the late 19th century is governed by the general system of settled cultivation ignoring the shifting mode. As a result the simple, free traditional cultivator is being forced into a new situation of master–subordinate relationship unknown in their system. The psychological adjustments which they had to make turned them into recluse and made them prone to depressions which manifest in their unwillingness to work, laziness, lack of skill as excuses. These depict the fact that transition is a super-imposed process which could not transform their economic structure totally and they got caught up in a vicious circle of shifting-plough-marginal farmers-agricultural/urban labour-shifting.

It ought to be remembered that shifting cultivation is tied up with the indigeneous life style, it is a mixed problem, partly connected with the social customs, traditions and beliefs. The present day shifting cultivator faces a big problem–neither can he become a full fledged plough cultivator, nor does he know the method of shifting properly, he is not even sufficiently skilled to work as construction labourer or get such opportunities too. So being aware of the unviability of shifting, denudation of forests, erosion of soil, he is unable to adapt to the changes. The pace of change is too fast compared to that of their community, the other facts such as religion, culture, society remains unchanged while the cultivator finds himself torn between the dualism of tradition and modernity. Some do try to match their steps with the prevailing change, but find themselves at losing ends often, as their resources do not sustain them for long and they fall into the debt trap ending up in a hand-to-mouth situation. The flat valleys got exhausted fast, slopes began to be terraced but lack of irrigation facility, inadequate capital investment acted as constraints. The administrative formalities in assisting the shifting cultivator to go through the phase of transition are a long drawn process and whatever help is given comes too late which forces the shifting cultivators to draw sustenance from elsewhere during the gestation period, and in this only those who are shifting cultivators-cum-settled ones survive and those with nothing to support them have to return to shifting once again.

Therefore, the discussion portrays the fact that the primary mode of agriculture in the state is going through a process of transition and the basic issue involved here is to make this process smooth and acceptable. Development schemes have already been introduced but it is seen in the course of study that they have not been able to wean the shifting cultivators totally. One factor comes clear, the shifting cultivators are not averse to the transition, they have already started developing permanent fields in the valleys and on the gentle slopes. Though these are the last resorts of the needy, yet it marks a qualitative change in the labour organisation and concept of property. However, land which can be converted into permanent fields in these areas is limited as also in this stage mutuality breaks down for the very reason that there is not much land available for permanent fields. And those who can avail of these permanent fields even acquire economic power and wield social influence much to the abhorrence of the community and threatens its structure on the whole. Shifting cannot be stopped completely, but at the same time it may be improved by better technology, seeds, fertilizers made available through governmental agencies. The hardcore jhumias deserted the colony scheme and went back to the hills, which put forth the failure of this scheme, at the same time, quite a few jhumia-cum-agricultural labourers stayed on and met with success. Establishing an egalitarian society similar to that of the past jhumia environment is not possible at the moment, they are not yet prepared to participate in labour-based asset creation and so developmental projects cannot benefit them. The pilot projects need to be so made that it portrays a clear policy and a thorough idea of these cultivators' way of life. There is no use eulogising shifting as a way of life, and that those involved are deeply concerned about its survival, its viability, land utility need to be studied with a firm determination. If a better know-how can be provided and several channels introduced for the betterment of forestry animal husbandry, land utility, tree-culture, the hard core jhumiyas can attain significantly higher level of achievement within a short period. Their traditions and beliefs need not be discarded outright as that will make the entire process of transition go haywards, the jhumiyas themselves will discard conventional values of life once they get past the transitionary phase. A full employment economy if introduced in villages would provide a viable economic base in due course. For this purpose the state administrative machinery has to act, removing the constraints, doing away with the deprivations of the shifting cultivators. Sudden change without adequate preparations will only put back the clock, rather a responsible body be entrusted with confidence implementation of these economic programmes thus removing the psychological barriers. Once the government acts in the right direction to help those concerned perhaps we will be able to find a new community of cultivators who have forsaken shifting and are the proud owners of settled land.

Field Study

In order to get a clear view of the transition taking place in the state economy regarding shifting cultivation, a field study was undertaken. A visit to the Southern district of the state revealed a number of purely autochthons' villages spread over in both hill and valley regions (Table 4.2). Shifting mode of cultivation came naturally to them but settled mode of cultivation too had acquired immense popularity. Two villages were chosen in the Belonia Sub division, Hetalia and West Paikhola which can be categorised as tribal villages. The inhabitants even claimed to have been 'Jhumias' or shifting cultivators or their descendants. While the Tripuri community inhabited Hetalia there were quite a few Reang families in West Paikhola along with a majority of Tripuris. As seen in Table 4.2, approximately 78 households in the Hetalia village and 41 households in the West Paikhola village were visited of which 43 households from the former and 19 households from the latter were picked up at random. The average household size was 5.17 and 5.09 per cent respectively of which able bodied men and women constituted 3.42 and 3.51 per cent who used to carry on shifting in the vicinity but owing to a constantly growing population pressure, extension of non-tribal habitation and lowering of yields in shifting cultivation, had to restrict their method and opt for settled cultivation occassionally reverting to shifting. Enquiries revealed that these autochthons were quite eager to take up plough cultivation rather than continue uneconomic shifting simply because of its traditional value on ritual bindings. They were quite aware of the low productivity of shifting, its harmful effects on environment and ecology, its seasonal nature as well as inadequacy in meeting their yearly needs.

Table 4.2: Sample Size Characteristics

		Hetalia	*West Paikhola*	*Total*
1.	Number of sample villages	1	1	2
2.	Approximate number of household in sample villages	78	41	119
3.	Number of sample household	43	19	62
4.	Average household size of sample	5.17	5.09	5.14
5.	Able bodiedmen/women	3.42	3.52	3.47

Source: Survey Data.

Table 4.3: Category-wise Distribution of Sample Households by mode of Cultivation

		Hetalia	*West Paikhola*	*Total*
1.	Exclusively practising shifting cultivation	5	1	6
2.	Exclusively practising settled cultivation	None	2	2
3.	Practising both shifting and settled cultivation	31	11	42
4.	Others	7	5	12

Other include non-cultivators like agricultural labourers, plantation workers etc.

Source : Survey Data

Table 4.4: Mode of Labour used in Cultivation

		Hetalia	*West Paikhola*	*Total*
1.	Using family labour	31	11	42
2.	Using hired labour	19	8	27
3.	Using exchange labour	7	3	10

Source: Survey Data.

In Table 4.3 it can be seen that out of a total of 43 sample households in Hetalia only 5 exclusively practise shifting cultivation and there are none who practice settled cultivation exclusively but a majority of 31 families practised both settled and shifting. However, quite a few households were categorised as non-cultivators which included wage labourers in agriculture and other allied jobs in plantations. The picture in West Paikhola village is not different. Here out of 19 sample households surveyed only one confessed of practising shifting cultivation solely, while 2 practised only settled cultivation and 11 households practised both settled and shifting cultivation leaving about 5 households which were non-cultivators engaged in plantation work or working as agricultural labourers. The survey conducted brought out certain facts which highlighted the transitionary phase. Farmers practising shifting cultivation exclusively in West Paikhola village related their woeful tales of having to practice shifting cultivation rather forcefully since their only privately owned land had to be sold off owing to indebtedness incurred during a social festival (marriage). The 5 families of Hetalia too had the same problems, which leads one to infer that shifting was not preferred but had to be resorted to in case of dire poverty. Though Hetalia did not have any exclusive settled cultivators, West Paikhola did possess two such families who were considered to be quite well-off financially according to the standards and are also revered to socially. Yet households practising both settled and shifting outnumbered the previous two categories. In Hetalia there were 31 households engaged in such a mixed cultivation mode and in West Paikhola 11, which accounts for a majority which portrays the phase of transition as settled cultivation is quite adhered to while shifting is not totally discarded. The idea one gets from here is that these two methods complement each other and this mix-up is much preferred by all the households. However, the presence of non-cultivating households reveal a sad picture of the villages. Enquiries show that these families did possess plots of land and practised settled cultivation only not being much adept at shifting cultivation. But gross indebtedness led them to dispossession of the land and not being able to fall back upon any other land (private or communal) turned to agricultural wage labourer and acquired jobs at nearby plantations. They too are of the opinion that after accumulating some assets they will be able to buy back their plots and embark upon settled cultivation once again.

Regarding the labour used here as the Table 4.4 portrays, two main categories are broadly deciphered, viz., family labour and exchange labour. Family labour can be understood as that given by the able-bodied men and women of a household. During peak seasons, family labour cannot cope with the work load and so exchange labour emerges. This had been the normal structure of labour in the past but with the introduction of the private property, traditional ties of reciprocity began to loosen and the kinship bonds, interpersonal contact weakened thus making it difficult to continue with exchange system. Meanwhile monetisation helped to sort out such complications and another category evolved and hiring of labour came into existence which in other words may be termed as exchange labour in a modified form. This labour structure could be seen in the surveyed villages which again portrayed the transitionary phase. In both the villages, Hetalia and West Paikhola, family labour was more prevalent as majority of the sample households (Table 4.4), 31 in Hetalia and 11 in West Paikhola use this mode. However, exchange labour seems to be on the wane as only 7 households in Hetalia and 3 households in West Paikhola use the exchange labour system. The reason here is already discussed, which mainly accrues to the nature of transition, alongwith the loosening of kinship ties the exchange system breaks down. Rather, the modified form of exchange labour has evolved and hiring of labour seems to be quite in vogue. Table No. 4.4 shows that 19 households in Hetalia and 3 households in West Paikhola have to depend upon hired labour off and on for their cultivation work, specially during the peak seasons. This even explains the presence of a non-cultivating category among the peasants. This examination of the labour use pattern, establishes a class-based society among the tribals alongwith the presence of a certain degree of economic and social difference among them, a direct consequence of this transitionary phase which was totally unheard of in the past among the shifting cultivators.

The field study, however, could not estimate the farm sizes properly. Hindrances such as the communal nature of possession of the jhum land, absence of such records, periodical dispossession of land due to indebtedness; rotation of the jhum land were faced and the holdings sizes could not be adjudged. As for those with settled farms, seemed slightly better off and even literate to some extent, could produce certain land records but it did not always match their plot of permanently cultivated land (changing of possession rights, break-up of families, partition, repayment of debts etc. can be given as the reasons for this). Though statistics were misleading, one pertinent assessment could be made here all the sample households had an inclination to possess valley/lunga land for settled cultivation and whenever some assets could be accumulated a piece of plot (for plough cultivation) was purchased irrespective of its being small, or even a marginal holding. The products of their cultivation generally hovered around rice and vegetables (pumpkin, gourds, potatoes etc.). They prefer marketing their products keeping a minimum for their self-consumption, another impact of the transition fanned by monetisation of the economy and the cultivators incline to a significant shift in favour of cash crops.

Yet the problems remain of indebtedness as their aspirations and the products never match, as could be understood from the survey. The tribals' abhorrence for colony life could also be felt as they thought colonies confine them within the artificial boundaries and hinder their independence. They seemed to prefer settled cultivation only but in their well acquainted environment, a direct effect of transition.

NOTES AND REFERENCE

1. F.A.O. Staff: *'Shifting cultivation'*. The Unasylva Vol. II. No. I, Rome Italy, 1957.
2. *Ibid.*
3. Jones, C.F. and Darkenwold. *Economic Geography*, New York, U.S.A. (1950), p. 143.
4. *Report of the Task Force on shifting cultivation* (1983) North Eastern Region (Census extract 1991, Govt. of Tripura).
5. Ramaiah P., *Tribal Economy of India*, Light and Life Publishers, New Delhi, India 1981, p. 32.
6. Goswami, P.C. and Boruah, D.A.: *Comparative study of crop production under shifting cultivation* (A case study of Garo Hills, Meghalaya), *Agro-Economic Research Centre*, Jorhat, Report of 1980, p. 11.
7. Conklin, H.C.: *The Study of Shifting Cultivation,* Honuroo Agriculture in Phillippines F.A.O. Rome, Italy, 1951 in current Authropology.
8. F.A.O. Staff: *op.cit.*
9. Ganguly, J.B.: *Economic problem of Jhumiyas of Tripura,* Bookland Pvt. Ltd., 1969, pp. 5-6.
10. To be precise 25,719 out of a total population of 74,242 practiced shifting mode. Hunter W.W.: *A Statistical Account of Bengal.* Vol. VI, 1876, p. 503.
11. *The Imperial Gazetteer of India,* Vol. XIII OUP London 1908.
12. Hunter: *op.cit.*, p. 503.
13. Deb Barma, S.C.–Census Bibarani (1340 T.E.) 1931, Agartala pp. 90-93.

14. Administrative Reports of the Tripura Govt. 1347 T.E. 1348 T.E. and 1349 T.E. sections 176, 177 and 178.

15. 16 drones 15 kanis = 108.4 acres.

    10 drones 3 kanis = 67.6 acres.

16. The Census Reports, 1931 (Extract)–Govt. of India.

17. Hunter, W.W.: *op.cit.*, p. 503 (when the jhumiyas found that jute thus produced can earn more cash than any other crops).

18. Administrative Reports of the Tripura Government - 1347, 1348, 1349 sections 176-178 (Govt. of Tripura), TSA.

19. Sharma, B.D.: *Shifting Cultivation and Their Development,* in V.S. Mahajan edited 'Studies in Indian Agriculture and Rural Development, Deep and Deep Publication, New Delhi, 1992, pp. 593-94.

20. Ganguly, J.B.: *op.cit.*, pp. 46-47.

21. *A bench mark survey conducted by the Tribal Welfare Department*, Govt. of Tripura, Agartala during 1978 showed the dependence of the shifting cultivators on money lenders.

22. *Ibid.*

23. *Report of the Commissioner of Scheduled Castes and Tribes*, 1955, Part 1, Pubd. by the Manager of Publications, Govt. of India, p. 123.

24. *Ibid.*

25. *Forest Rules of Tripura*, Tripura Administration, 1957 Rule 3 of Notification No. 3, Govt. of Tripura, p. 4.

26. *Reports of the Planning Commission–Plan,* Outlay (Five Year Plans) VI, Plan, Govt. of India, 19.

27. Project Reports of the Tripura Rubber Plantation Corporation.

5

Ziratia Land

The hilly terrains of the state seemed ideal for shifting cultivation and the indigenous people depended on this activity primarily for their livelihood. But there were ample plainland, lush green valleys throughout the state which lay waste as plough cultivation did not gain importance. Shifting cultivation, on its part could not yield any surplus or for that matter did not add to the state revenue even. As the British clamoured for enhanced returns the kings had no other option but to rely on land revenue which constituted quite an important source of revenue. So, in their quest for increasing the land revenue, the kings decided upon various ways of leasing out land. Fertile, plain land, lying abandoned where plough (which could fetch a substantial amount of revenue) could be carried on were given out on leases. Attracted by such proposals a few permanent residents of the neighbouring British Bengal came over to the state for cultivation purpose solely. They were not permanent settlers of the state and resided in British Bengal side near the borders (on the East Pakistan side). These cultivators were known as Ziratia peasants, the word originating from the Bengali word 'Zami-Zirat' meaning agricultural land.[1] The present chapter deals with the land of these cultivators, typical of Tripura, how the cultivators came, whether they 'owned' the land and then what happened to these lands once the peasants returned to their home (erstwhile East Pakistan) after the partition.

From the very beginning of British rule in Bengal, land revenue exactions were exorbitantly high. The landlords had to depend upon their tenants or lease holders to pay this revenue, hence the rentals imposed on the latter were high too. Paying such huge exactions the peasants remained perpetually impoverished, even paying by free labour at times. The tenure of their holdings was also quite insecure being merely tenants-at-will. During the mid 19th century when the big landlords of British Bengal increased the rentals of their states and pressurized their tenant-cultivators to pay more rents, the latter could not meet such excessive demands as the rent were beyond their capacity neither could they press their customary rights in courts working under a constant threat of eviction as they cultivated land on lease. Consequently, these peasant cultivators faced grave economic hardship which drove them into the clutches of money-lenders for payment of rents as well as subsistence.[2] Indebtedness led to frequent land transfers from the peasants to the money-lenders or big landlords which was accelerated during natural calamities (floods, famines, pestilence etc.) and crop failures. To quote Bipan Chandra "...The peasantry was crushed under the triple burden of the government, the Zamindar or landlord and the money-lender".[3] The peasants were out of work, relinquishing their hold on land and contemporary estimates show figures ranging from a third and four-fifth of the total land remained uncultivated.[4] At this juncture it was not land which was scarce,[5] but tenants who would cultivate land in British Bengal. There were others who refused to pay rent and so had to abandon their lands. Those farmers who could survive such calamities ultimately ended up as landless labourers being compelled to sell their products immediately after harvests at low prices to the merchant-money-lenders in order to repay debts, and even had to dispose of their land to buy other necessities. At this point of crisis, owing to geographical proximity, those residing in the bordering areas of British-Bengal decided to venture into Tripura in search of livelihood.

Now, such practice was not totally unknown to the people residing on either side of the state. The autochthons of Tripura, generally averse to change over from shifting to plough, found that certain hill-men took up waste land in the plains near their villages and cultivated it through some Bengali-Muslim peasants whom they employed as servants.[6] The kings of Tripura had a Zamindari on the fringes of the state, Chakla Roshanabad,[7] which rendered a substantial amount of revenue and was cultivated by peasants who were Bengalis residing in British Bengal. These peasants depended greatly on the Tripura State economically and the altered political relations after partition did not affect the relation between landlord and the tenant. As the landless peasant-cultivators turned towards the east, they found Tripura to be a haven of agricultural land as it was easily available. Almost during the same time, the king of Tripura too needed more peasants who could take over the plough cultivation and enhance the state revenue. Hence various concessions were declared like the Jangal Abadi Bandobasto[8] order where in the king gave a open call to the tenants of British Bengal to come and clear the jungles for extensive cultivation. Not only this, even the rates of rent were kept low (ranging from 0.62 paise per acre in the remote acre as to Rs. 4.37 paisa per acre in plainsland while that of Bengal was Rs. 6.18 per acre)[9]. The land on the border areas near Noakhali and Sylhet were lying abandoned and these were quite fertile, hence most attractive to the landless cultivators of British Bengal.

These peasants, mostly Bengali Muslims, flocked in hundreds and since the cultivators did not reside within the territory of the princely state of Tripura permanently but returned after harvests carrying with them the crops produced, they came to be known as Ziratia peasants. These peasants assumed great importance in due course of time in the agrarian economy of the plainlands of the state. They also helped in developing the resources of the state as they were very hard working and even brought jungles and uncultivable land under intensive cultivation. But once the gates were opened wide for their entry, their numbers multiplied within years which resulted in another mad scramble for land and ownership rights. Though they were not residents of the state strictly they were welcome to settle in any part of the border area on payment of a nominal rent (adda, a type of chowkidari tax realised from settled cultivators only) to the state. Presence of too many Ziratia peasants resulted in cheap labour and scarcity of arable land leading to creation of hierarchy of owners as sub-letting, sale, transfer and mortgaging of raiyati interests led to fragmentation of land within a short span of time. Yet since they were able to bring extensive land under cultivation as large tracts of land lay fallow the kings of Tripura too encouraged their settlement with the modernisation process. Although the Ziratia peasants had land holding in vast areas along the western border of the state, Sonamura division had the largest concentration followed by Belonia district. Even B.C. Dutta mentions that the Raja's subjects in the plains of this region (Sonamura) were Ziratia Muslim tenants.[10]

The coming of these non-resident Muslim cultivators from the neighbouring country initially did have some positive results. Certain rules regarding their cultivation were imposed, they were required to obtain permits to reap their own crops; the paddy sent out of the state was liable to an export tax of Rs. 15/- per maund which they had to clear to reap aman crops, they had to supply 10 seers of paddy to the State's treasury.[11] These peasants who were either landless or nominal landholders in British-Bengal prospered highly by carrying on cultivation in Tripura as the reports reveal in 1931, 26.19 lakh maunds of rice amounting to Rs. 52.38 lakhs were exported from Tripura during the Ziratias' annual homegoing, while at the same time 12,000 maunds of rice had to be imported.[12] From the figures it may be understood that the coming of these Ziratia peasants could not help the state treasury to boost up actually, whatever crops were produced were transported to their home country while Tripura state remained in the clutches of crop failures. It was regarding this question that controversies arose between the state administration and its British-Indian counterpart. As the state had to import quite a large amount of rice during this time, objections were raised about the lenient terms meted out to the tenants, and an embargo was imposed on the export of rice from Tripura by the Ziratia peasants henceforth.[13] Obviously, the embargo was disliked by the cultivators of British-Bengal and they wanted to take away the entire paddy grown by them for their own consumption. They were backed by their government too. Though land available for plough cultivation and a rich source of crop production as well as revenue generation was lost, the relation between the king and the British-Indian government was not much affected as the former virtually surrendered to the latter's demands, and this understanding went on smoothly for some years.

The partition of India and the subsequent merger of the princely state of Tripura with the Indian Union (1947, 1949) considerably altered the existing pattern of relations. The upheavals which followed the political changes were given a religious hue and these soon strained the relation between the two sides. As harvest time drew near, the Ziratia cultivators dared not enter Tripura to reap the paddy grown by them as cordial trade relations too worsened.[14] Gradually such chaos and tension was controlled and the Inter-Dominion Agreement passed an order limiting the export of paddy from the state to 20 maunds per family, subject to payment of export duty at the rate of Rs. 1-2-3 pies,[15] and the rest of the produce had to be disposed of in the state. Now, this would perhaps lead to a gain for the state government which have been forgoing quite a substantial amount of crop produced by the Ziratia peasants due to the peculiar understanding made by the kings in the 19th century. But the peasants were obviously perturbed as they did not want to forgo any of their privileges and so gave vent to their grievances. With the introduction of the Visa and Passport system in 1952,[16] the Ziratia question took a new turn. The tenants' movements were restricted, they were not allowed to cultivate the lands which they all along thought to be their own, the duties, rates of tax (adda) were enhanced too. Tension brewed on both sides and some peasants even claimed that all arable lands of Hill Tipperah was under the ownership of the Pakistani border people, they went to the extent of forceably taking away a great deal of rice without any valid permit.[17] Rumours spread that the Ziratia peasants were coming in large numbers with the aim of settling here permanently.[18] However, there was marked decline in the number of operational holdings of the Ziratia peasants and the area too. On partition there were 20,630 holdings with a total land area of 29,078.13 acres which decreased to 13,014 with a land area of 18,494.80 acres.[19] This shows that instead of coming en masse the Ziratia peasants decided to return to their original homeland, abandoning their plots in Tripura. Though there were about 889

Ziratia peasants of Indian origin cultivating land in East Pakistan, they were neither given any concessions nor pampered like their counterparts in the borders of Tripura.

However, one point was crystal clear, the Ziratia peasants being mostly Muslims, were unlikely to stay in India fearing an insecure future. But the question of whom to sell their land remained unanswered as the Government of India initially failed to adopt any definite policy on land owned by them.[20] A policy of serving certificate notices on the absentee Ziratia peasant was undertaken so that they pay the rent of their plots in accordance to Section 32 of the Tripura Sarkari Prapya Adaya Sambandhiya Ain, Act 4 of 1326 T.E. and if they did not show up, the rent was to be realised from the sale proceeds of paddy procured from their plots. Quite an extensive region was lying abandoned after the political turnmoil following partition and therefore, the Collector, under Arable land (Utilisation for food crops) Order, 1950, had the power to call upon any peasant (holder of a plot or not) to cultivate such land by himself or through tenants with any food crop and if the cultivator fails to comply, the Collector can give the plot of land on lease to the person he deems fit, for a period of 3 years. Though the order was passed by the government it could not be enforced on the territory. Land which belonged to the Ziratia peasants now was being taken back for developing the state's own economy by boosting crop production, but problems persisted. Those who inhabited these regions (specially the displaced) were unwilling to take lease or cultivate the land unless the Ziratias could be adequately compensated, perhaps they feared reprisals as the land was on the borders within easy reach of the Pakistani nationals. Secondly, these regions now were under thick jungle once again which would entail considerable initial expenditure by the persons who try to cultivate it. Thirdly, once the person take lease of some land he is likely to build a house on it and when called upon to vacate the same after 3-4 years this house would stand on the way.[21] Rather, if land would be given on raiyati rights, then cultivators might come forward, or even if given through cooperative societies, these lands would not remain fallow for so long. Quite a considerable portion of the land was acquired by the mid 1950s (42 drones, 14 kanis, 15 gandas, 3 karas) compensations awarded even but could not be paid to the peasants who were Pakistani nationals. A proposal was made to the Government of India that the entire length of Ziratia land be acquired for rehabilitating the displaced persons from East Pakistan.[22]

Analysing the foregoing facts it appears that a lot of complicacy was created about this aspects of land question. Sonamura, as already stated, had the largest concentration of Ziratia tenants and the problem was most noticeable here. The Sub-Divisional Officer of this division prohibited anyone, who had purchased such land even 15 years ago, from harvesting if they could not obtain a permit to do so.[23] This bit needs to be explained here, most of the Ziratia peasants decided to return to Pakistan and left their land to their labourers (Hindus) on minimum prices without any lawful land transfer. Later, a novel system of exchanging of property was evolved to facilitate the people residing on the borders mostly. Those Hindus who resided on the borders of erstwhile East Pakistan acquired land and even citizenship of India through exchanging of properties by Power of Attorney. But they were merely land-holders on the other side who came over to this side as land-holders again, not on being displaced from their original homeland. Thus, these, border land could not be turned into Khas land or used for any public purpose even as quite a considerable area still belonged to absentee Ziratia peasants.[24] Now this role of the government of letting matters take its own course was much abhorred by the peasants as reported by the All India Kisan Sabha Union.[25] The question of Indian ownership of lands of the Pakistani Ziratia tenants remained unresolved, they had to wait will mutation to get hold of the land. An attempt was made in the 1950s to ease the problem of the land of the Ziratias and distribute the same among the people of the state. Land was categorised in accordance to their ownership rights of distribution[26] land purchased by Indian nationals but not yet mutated; Ziratia land cultivated by the Indian nationals as bargadars (sharecroppers); land cultivated by the Ziratia tenants themselves. Former rules and regulations were nullified and new rule implemented regarding harvesting. While those possessing the first category land could harvest their crops on producing documentary evidence, those with the second category land had to sell half the share of harvest to the government at fixed price and the Ziratia peasants were required to collect and sell the entire harvested crop to the government and get his share of paddy. Naturally such regulations regarding land ownership as well as harvesting did not find enough takers, even the district officials with whom was vested the power of distribution of such land, found it very difficult to enforce them. As a result vast expanses of such border land lay abandoned with thick jungle growth deterring those hundreds of landless peasants from cultivating it in face of dire land scarcity throughout the state. To locate the landowners and then take possession of their land after mutation simply seemed to be a herculean job for the state administration and this can be seen from the Table 5.1 which depicts the number of pending cases awaiting disposal and the number of applications received during May and August 1958.

Table 5.1 reveals a number of facts regarding the Ziratia tenants land problem in the post Independence period. That Sonamura division had the most number of such tenants can be seen at a glance but even after a number of years since Independence, the number of pending cases decreased however a large number remained and the ensuing problems can be well understood. Belonia division put forth a similar picture though the numbers were far less (almost half the previous ones), however, pending cases at the end of the time period remained almost unchanged, that means not much Ziratia land could be mutated. In Sabroom, the numbers were very few though the number of cases disposed during the said months were negligible. The western borders like Khowai or Sadar did not have as many Ziratia tenants, hence the land area affected was almost nil. There were no Ziratia land in Kailashahar, Udaipur and Amarpur divisions. The administration, undoubtedly, tried to solve this entangled problem of the Ziratia tenants land as is evident from a number of reports (unpublished) of the concerned departments, lying at the Secretariat Archives. They admit that the Ziratia tenants are not regular in submitting the necessary documentary proof in support of their cases and these are often dragged due to non-appearance of the applicants and non-payment of the 'nazrana' as required, or even non-production of evidence of possession.

Table 5.1: Ziratia Cases Awaiting Disposal (1958)

Names of the Division	*Year 1958*	*Pending cases (Nos.)*	*Application received (Nos.)*	*Total*	*Cases disposed*	*Pending cases at the end*
1. Sonamura	May	1668	33	1701	21	1680
	June	1680	98	1778	67	1711
	July	1711	256	1967	176	1789
	Aug	1781	244	2025	33	1992
2. Belonia	May	766	44	810	–	810
	June	810	55	865	5	860
	July	860	114	974	61	913
	Aug.	913	15	928	34	894
3. Sabroom	May	20	–	20	–	20
	June	20	–	20	–	20
	July	20	–	20	–	20
	Aug.	20	16	36	3	33
4. Khowai	May	3	–	3	–	3
	Aug.	–				
5. Sadar	May	2	–	–	–	2
	Aug.					

Source : Unpublished Reports of the Tripura Administrations Office of the Revenue Department, 1958.

In view of several complications of border area security measures, all the old rights of the Ziratia tenants had to be derecognised and this brought a lot of other problems e.g. trespassing, forceable entering to take away crop produces, forest products etc. The long international border of the state has been a source of lot of disturbances acting as a great hindrance to its development. The bountiful, fertile valley land which the Ziratia tenants once claimed to be their own did never come in the possession of the government directly, rather changed hands under covers during political upheavals in the two countries. These lands cannot be recorded as 'khas' (government owned) until it is declared so through a proceeding for arrear for land revenue.

Where such land is not possessed by any landholder the land will be vacant and in some cases leased to bargadars with the Sub-divisional officer's permission.[27] No such provision was made defining the status of Ziratia tenants in the Tripura Land Revenue and Land Reforms Act of 1960.[28] It simply provided raiyati status for such type of holder and thus the word "Ziratia" had to be deleted from the record of rights and substituted with the word "Raiyati".

Thus we find the Ziratia tenants land to be ultimately in the possession of Indian citizens who crossed the borders during the 1940s, 1950, 1960s and settled on the land not only cultivating the fertile land but also adding to the population pressure of the state. A study of the Ziratia land of Belonia division has been made which reveals certain interesting facts about the present owners of the land.

Field Study

The Ziratia tenants of Tripura was an unique feature in the economic history of the state. They did not add to the growing population pressure of the state undoubtedly but though they carried on cultivation on the land of the state, they did not contribute much to the state income since they initially took away all that they produced. This was quite a complicated issue, and to get an insight to it, a field study was conducted in the Belonia sub-division where quite a vast region was marked as Ziratia land. Villages like Barapathari, Manaipathar, Sonapur, Durgapur, Ekinpur, Ishanchandranagar, Ballamukha (with a Hindu majority population at present) Samarendranagar, Amjadnagar, Rajnagar (with a Muslim majority population at present) flourished during the early 20th century as ziratia tenants with their hardwork brought them under cultivation. 33 families including both Hindu and Muslims were surveyed from these villages and a study was made regarding the ziratia land.

Almost all of these families came over to this side of the state during the 1950s and 1960s to settle here permanently. The Muslim families are generally descendents of the ziratia tenants themselves who acquired the land through certain formalities with the government. The Hindus are mostly agriculturists who acquired such land on exchange basis as mentioned earlier though a few did admit that they had to resort to benami transfer of land as the former owners were in a haste to leave the land and they too wanted to settle here. However, most families claimed that they came here to settle only on being declared legitimate citizen of this country. The plots they had acquired were quite large and so land could be put to optimum use and these settlers were quite adept at ploughing too.

A few details regarding the former ziratia tenants was gathered from a senior citizen, the only living member who was originally a labourer of a ziratia peasant. Around 1935-36 a large number of such peasants turned up (during that time a scion of the royal family, Narendra Kishore Karta was the Talukdar of the region) as tenants, they were either landless or nominal landholders of British Bengal, but could not earn a decent livelihood there. These tenants were mostly Muslims while they brought along some Hindus too who could assist them. It was also learnt that geographical proximity between the two states made them return to their homeland every evening after a hard day's work. They used the fertile plain land for cultivation of paddy, jute while the hilly part (tilla land) was used as temporary farmhouses. Some even had orchards along their plots. These tenants were expected to pay a Ghasuri tax[30] (grazing tax) and also a nominal rent 'Adda' which was collected on a special day (Punya).

However, after Independence and following the partition certain changes were noticed. Many crossed the border to this side of the state in order to escape the atrocities inflicted on them and longed for a peaceful, secure future. Others crossed over to settle for further studies and employment and all were gladly welcomed to Tripura and pampered too. As the government decided to change the ziratia land to raiyati ownership many peasants found abundant cultivable land and could soon manage mutation proceedings to take over the plots. These fertile tracts were tilled by tenant-cultivators or share-croppers and yielded quite a large quantity of paddy for individual consumption or even markets. It should be mentioned here that on visiting these lands several plots were found lying vacant and uncultivated too, even the settlements were found abandoned. On making queries with the people it was learnt that of late there had been a tremendous rise in border crimes (dacoity, smuggling etc.) which made living on these fringes extremely precarious. Moreover, these second generation land-holders do not completely depend upon land for a livelihood, they are educated and prefer white collared jobs and insist on giving away their land on barga. But even these sharecroppers abandoned the land fearing insecurity of life. Hence the land which was procured from the possession of non-resident farmers to balance the pressure of population on this side of the border, seem to get lost in the turmoil of security measures.

It may be deduced from the discussion that the long international border between the two countries has stood witness to several changes, the ziratias were undoubtedly hardworking peasants who, with their hardwork did bring miles of uncultivated land under cultivation and produce quite a huge amount of paddy, but this had almost no impact on the state economy as they took away a giant share of their produce. However, since Independence this land was brought under cultivation by the Indian nationals, yet owing to increase of crimes in the area once again the land remains abandoned. The only answer for this complicacy is perhaps, strengthening of security measures and re-allocation of land through proper governmental machinery.

NOTES AND REFERENCES

1. Chakraborty, Mahadev: *The Ziratia Problem of Tripura,* Proceedings North-east India History Association, Fifth Session Aizwal, Mizoram, (1984).
2. Ganguly, J.B.: *The Benign Hills,* Tripura Darpan Agartala, 1983, pp. 20-24.
3. Chandra Bipan: *Modern India, National Council of Educational Research and Training,* New Delhi, p. 194.
4. *Report of the Land Revenue Commission,* Bengal (Chairman-F.L.C. Floud) 6 vols. Alipore B.G. Press, 1940.
5. Chatterjee, P: *The Land Question of Bengal* 1920-1947, vol. 1, K.P. Bagchi & Co. Calcutta, 1984, pp. 14-15.

6. *Annual Report of the Political Agent,* Hill Tipperah No. 149 dtd Agartala 31st Oct., 1873 in D.K. Choudhuri (ed) Annual Administrative Report of the Political Agency, Hill Tipperah Vol. 1, Tripura State Tribal Cultural Research Institute and Museum, Government of Tripura, 1996, pp. 17-21.
7. The Chakla Roshanabad Zamindari of the Manikya Kings comprised an area of 555 square miles of land and included 53 parganas, stretching along the whole length of Tipperah district (Noakhali in the South to Sylhet in the North). The kings derived a considerable amount of revenue from this plainland estate which contributed to the property of the state. In 1765, at the time when the East India company controlled the District of Tipperah the king had to pay rich tributes to the British. But owing to certain disputes over revenue assessment between the resident of Tripura, Mr. Ralph Leak and the king Krishna Manikya, the zamindari of Chakla Roshanabad was converted in to 'khas' i.e. taken under direct management, for ten years (1782-1792). At the time of the permanent settlement (1793) the estate was settled permanently with the Maharaja of Tripura at an annual revenue. After the merger of the princely state of Tripura there was no mention of the status of this royal estate and it was awarded to Pakistan by the Radcliffe commission.
8. Hunter, W.W.: *Statistical Account of Bengal* Vol. VI (reprinted 1973), p. 505; (Order No. 1293 T.E.) 1302 T.E. in D.C. Dutta and S. Bandopadhyay. Rajgi Tripura Sarkari Bangla. Directorate of Education, Government of Tripura, Agartala).
9. Report of Land Revenue Commission *op.cit.* Vol. II. Appendix IX Table VI (a) pp. 100-104.
10. Dutta, Brajendra Chandra: *Tripura Rajye Tris Batsar* (in Bengali) Sonamura Bibhag, Directorate of Education, Govt. of Tripura, Agartala p. 8.
11. Letter No. D-514 W-3-34/43 dtd. 11.12.1943, State Archives, Govt. of Tripura, Agartala.
12. Deb Barman, S.C.: *Census Bibarani,* Census Office (1932) Tripura State, Agartala, p. 108.
13. This is evident from an unpublished letter written by the Secretary to the Resident of for Eastern States to the then Chief Minister of Tripura State.

    Eastern States Agency No. W. 3-77/42-4571 from Capt. A.C.K. Maunsell, 30th Apri,l 1943 State Archives, Tripura.
14. Hindustan Standard, Calcutta dated 25.12.1948 published a news item -

    "...meanwhile reports about atrocities, extortion, molestation and humiliation committed by Tripura state police, specially of Sonamura police station and some miscreants on innocent Pakistani people particularly members of the majority community continued to pour into Pakistan.... All traffic to and from the Tripura State is practically at standstill....on account of prevailing tension and posting of military Pakistani cultivators dare not enter state territory at present to reap paddy grown by them."

15. B.No. 4/5 No. 25/1963. Letter No. 339, 11-11/53p. Secretariat Archives, Govt. of Tripura, Agartala.

16. An agreement was arrived at between the District Magistrate and the Superintendent of Police, Tripura on one side and the District Magistrate, Tippera and Noakhali on the other on 5 December, 1952 when the ziratias were entitled to 'A' category visas which could be issued by the District Magistrate.

    B 26/531/1952/FN–XX–113/52, Secretariat Archives, Agartala letter No.64-p/xx-113/52 dt. 5.5.1953.

17. They were allowed to take away 40 maunds of rice or the entire produce whichever was less according to the Indo-Pak Trade Agreement, 1955.

18. Unstarred question in the Rajya Sabha No. 21.18 February, 1958-answered by the Prime Minister.

19. Tripura Administration, Office of the District Magistrate and Collector (Revenue Deptt.) No. 12939/DM/REV/XI-3/59 the 11-11-1959/20-7-1881 (State Archives).

20. Cut motion in the Lok Sabha S.N. 1347 Demand No. 161, List No. 44, Demands for Grants in respect of General Budget, 1957-58.

21. Revenue Department *op.cit.* F.No. 12939/DM/REV/XI-3/59 11-11-1959/20-07-1881.

22. *Ibid* (The proposal was even considered in the course of inter Ministerial discussions prior to the Indo-Pak Trade talks early in 1957, but in view of the possibility of adverse diplomatic repurcussions, it was considered unnecessary to pursue the matter).

23. Tripura Administration office of the District Magistrate (Food Section) No. 21746/DM/FP/XVI-9/88 dtd. 8.12.58 Secretariat Archives Agartala.

24. *Ibid.*

25. Representation of the All India Kisan Sabha B/241/517/F.N. 2 (21)/-p/58 Secretariat Archives, Agartala dtd. 24.10.1954.

26. Evidence from the F.No. 21746 of the office of the District Magistrate (Food section), *op.cit.*

27. General Order No. 3 Technical Recording of land deserted by Bangladesh Nationals. No. 8250-53/F. 1 (58)-DLRS/ESTT/78. Directorate of Land Records and Settlement, dtd. 5 Dec., 1978.

28. Tripura Land Revenue and land Reform Act of 1960 (No. 43 of 1960) Govt. of Tripura, Directorate of Land Records and Settlement, Agartala: Complication of various orders, Memorandaum, Instructions issued on Revision of Record of Rights General Order No. 6: F.No. 1601-07/F. 1 (58)-DLRS/LR/78 dated 9th Feb. 1979, stated that Bangladesh Nationals holding the land within this state was recorded as 'Jiratia tenants since ownership of immovable properties in India by the other nationals was not barred by any law in force. But there is no provision defining the states of such type of holders. The word 'Jiratia' be deleted and changed to Rayati.
29. General order No. 6. Technical No. 1601-07/F 1 (58)-DLRS/LR/78 dtd. 9 Feb., 1979.
30. By an order dated 7th Ashar, 1358 T.E. the Tripura Gazete passed the following order 'No grazing fees should be charged on any cattle if it grazed on the owner's own settled land. But the ziratia tenants were being denied this benefit as they were not allowed to take their cattle over the lands of others to graze in their own land, and so had to pay the Ghasuri tax (from a discussion with the original ziratia labourers during the course of the field study.).

6

Impact of Forest Conservation

Land acts as the foundation of a country's economic setup and forests have a definite claim on this resource. Forests are considered to be the essence of environment and are essential for the maintenance of the physical characteristics of the land. Soil conservation, a necessary prerequisite for agriculture in scientific lines is ably done by forests by preventing the silting up of reservoirs in catchment areas.[1] They help in increasing the fertility of arable land and improve their moisture content too, besides providing shelter and seclusion for those who inhabit them. On man's gradual entry to the phase of agriculture, forests fostered the human society with care and affection by preserving soil, providing fuel wood, coal and timber. With ample land and sparse population during the past, shifting mode of cultivation was pursued widely with little or no impact on the forests. As settled cultivation was introduced man's dependence on forests declined but land had to be cleared and forests receded in direct proportion to the expansion of plough. Population began to increase those depending on shifting was affected, their land grew scarce fallow cycles shortened and the natural equilibrium seemed to be threatened. Indiscriminate felling of trees made the importance of forests lie only in folklores, rituals and ceremonies. Developmental activities proved hazardous to the environment as denudation set in.

On the Indian scene, the colonial British rulers never gave much importance to the forests during the last century. In fact, they tried to siphon off the resources of the forest land to serve their own commercial interests. They did not even bother to introduce improved technologies so that the basic needs of the population could be met with reduced resources.[2] The big landlords owned the forest land; but with the abolition of the "Zamindari system" land changed hands and got increasingly subjected to depletion as encroachments became rampant. Hence forestry oriented activities had to be undertaken by optimum use of land resources and available man-power, but a pertinent question was raised here-whether to preserve forests or rehabilitate the people. Forests on the whole, in the national scenario, lay claim to an adequate share of land and its importance can hardly be over-emphasised in a region where agriculture constituted the predominant livelihood of the people. The present chapter focusses on the role of forestry as a vital source of wealth and the impact of the Forest Conservation (Acts) on the land use pattern of the state of Tripura. A quick perusal of the forest cover in the North-east region acts as a backdrop to the topic.

The North-Eastern sector which covers a total of 25.50 million hectares of the country has a large forest cover too. According to the National Remote Sensing Agency's estimate report (1983) based on satellite imagery (1972-75) and (1980-82) the forests cover 48.28 per cent of the total area.[3] Of this area again it may be stated that forests largely remain unutilised and unexploited as the lie outside the direct control of the state, managed by forest dwellers who enjoy historical prerogatives being inhabitants of the forests since long. Forests of the region provide a number of products for household use (firewood, charcoal, fodder, essential oils, honey, lac etc.) or industrial use even (paper, matches, rayon, construction materials etc.). However forest land in this part of the country is subjected to much depletion, shifting cultivation and cutting down of trees to create land for agriculture, which is more profitable, being the main cause for the same. Due to absence of clear cut policies regarding forest land, survey and settlement operations could not carried out in most of the north-east region which acts as a serious impediment in embarking upon a forestry programme. North-eastern tribes are quite dependent upon the forests but pressure of population, environmental requirements calls for preservation of this land mass too. Exploitation has to be controlled and at the same time quick growing species, plantations need to be grown. Merely classifying forests according to Reserved and Protected regions do not serve this need.

The Table 6.1 clearly depicts the total increase/decrease in the forest cover during the past decade in the region. All the states have maintained the same cover area during the period except for Nagaland which shows an increase of 575.3 and Tripura 38.7 (in 000'ha). Astonishingly, the two states with highest immigration problems reveal opposite characteristic in forest cover, while almost 60 per cent land of Tripura is forest, less than 50 per cent of Assam has forest cover. One point comes clear that there is a general tendency of depletion (loss) in the forest cover of the region as estimated by Forest Survey of India[4] approximately 635 sq. km While only Nagaland and Tripura have added (27 sq. kms. and 3 sq. km. respectively) all the other states have shown remarkable depletion (mostly because of shifting cultivation) despite regeneration in abandoned areas.

Tripura, during the previous centuries was covered with dense forests, but neither the kings nor the British authorities in the later years paid any heed to them. It was left to the autochthons of the state, who carried on shifting cultivation in the forests as well as enjoyed the products by themselves. However, the scene changed with the opening up of the state to British intrusion during late 19th century. New markets were opened in the neighbouring regions of Chittagong, Noakhali, Sylhet[5] and the importance of forest products was increasingly felt, and there started a systematic exploitation of the rich reserves. The forests were looked upon as the most important source of revenue belonging to the state. Even the elephant 'kheda' operations i.e. royalty on elephants caught fetched a lumpsome as the Mughals were attracted for this previously. These too were given on ijaras to those who paid the state a share of the value of elephants captured varying from 2 annas to 7 annas share. The forests were also leased out on the farming system as 'mahals' at absurdly low rents. There were as many as 28 farmers of forest produce as late as 1874-75 whose

Table 6.1: Forest cover of North-east Region, 1980-85

('000 hectres)

State	Year	Geographical area	Forest area	Reserved forest	Protected forest	Uncleased forest	Owned by forest deptt.	Owned by individuals
1. Arunachal	80-81	8374	5154.0	1183.0	20.6	3935.5	5127.5	14.7
Pradesh	84-85	8374	5154.0	1337.0	0.8	3730.9	1323.1	14.7
2. Assam	80-81	7844	3070.8	1707.4	–	275.9	1777.3	–
	84-85	7844	3070.8	1727.7	337.3	1005.8	1745.9	–
3. Manipur	80-81	2233	1515.4	137.7	417.1	960.6	1515.4	–
	84-85	2233	1515.4	137.7	417.1	960.6	1515.4	–
4. Meghalaya	80-81	2243	851.0	70.6	1.2	779.2	71.8	779.2
	84-85	2243	851.4	71.0	1.2	779.2	72.22	779.2
5. Mizoram	80-81	2108	1593.5	561.8	164.7	524.0	712.7	–
	84-85	2108	1593.5	904.8	164.7	524.0	712.7	–
6. Nagaland	80-81	1658	288.2	46.7	51.8	–	46.7	189.7
	84-85	1658	862.5	100.4	–	762.1	100.4	762.1
7. Tripura	80-81	1049	592.2	386.5	205.7	–	599.2	–
	84-85	1049	630.9	386.3	–	244.6	630.9	–

Source : Ministry of Forest and Environment, Government of India, Basic Statistics of the North–Eastern Region, 1995.

rights extended only to collecting of forest dues of produces removed by land or water.[6] Tolls were levied by the kings of the state of which the British government claimed a 6 anna share.[7] At times the management of some of the forests were taken up by the ruling class and this increased the revenue returns.

Meanwhile, the British government passed the first Forest Act in 1865 to ensure regular returns from forest revenue and created a forest department too.[8] In 1871 the Chittagong Hill Tracts were declared government forests but with inaccurate knowledge and absurdly low rents, no systematic trade in jungle products could be carried on here.[9] In 1885, the then ruler of Tripura Maharaja Bir Chandra Manikya introduced the concept of state ownership and the forests were classified into 'reserved' and 'protected' categories as their revenue generating capacity was realised.[10] In 1886, was passed the first Act relating to the forests of Tripura. The Act denied the rights of the forest dwellers to 'elephant trapping operations' or cutting down of various valuable trees (Sal, Garjan etc.) as the state was the lawful owner of this resource. The next year (1887) the king went a step further, when he declared shifting cultivation to be completely prohibited in the forest regions.[11] This procedure of increasing revenue from the forests since settled cultivation was not yet widespread was gaining ground in revenue administration, the kings even thought of novel techniques to enhance the same. Thus started a new scheme, 'Jangal Abadi' (1889).[12] According to this scheme, land was to be kept rent free for a period of 4-10 years, subjective to the nature of the forest. It was also to be ascertained whether 'muddat-minah' can be given i.e. till a definite time (Muddat) no rent will be charged but after that time period rent is to be paid in accordance to the quantity of land cultivated every year. The rate initially hovered around 2-3 annas (12-18 paise) per bigha which later increased gradually till it reached a moderate rent.[13] It needs to be mentioned here that even this rate was not equal to the rent of the adjoining plainsland in the government territory. Therefore, the scheme attracted many farmers and talukdars from the neighbouring plainsland, more so since shifting was being prohibited. Forest territories were cleared at the edges to build fresh settlements which meant more and more land would be available for plough cultivation. Once 'settled' this population added considerably to the cultivating population of the state leading to a new aspect of population explosion.

In order to preserve the forests from onslaughts of uneconomic and wasteful shifting, the Manikya rulers were strict with the regulations. Any infringement of such regulations would be dealt with severely according to a Forest Rule of 1903.[14] In 1908-09, a total of 20 sq. miles area was declared as reserved and 3861 sq. miles as 'unclassified open' forest in order to protect the forest land. In this context, 'reserved' meant no one was allowed to 'use' forest produce, but this settlement could not continue for long. Income from Kheda operations (royalty on elephants caught) was gradually falling during this time. A plausible explanation is reclaimation of marshes and extension of cultivation and clearing of forests. The deprived forest dwellers began to agitate and clamoured for reinstatement of their rights on the forest products, and therefore by an official promulgation in 1913 emphasis was laid on reserving 'trees' rather then 'areas'.[15]

Forest Conservation (Acts) and Land-use Pattern

The forest dwellers initially depended upon shifting in the forests during the 19th century when there were abundant land available. With growing population pressure in the early days this shifting 'cycle' was reduced comparatively and forest resources declined in direct proportion leaving mass destruction in these once luxurient forests. During lean seasons, these forest dwellers had to depend upon forest produce to draw sustenance, but with continuous destruction such dependence declined (e.g. collection of fuel wood, fruits, dryleaves, felling trees for timber etc.). Their traditions were woven around the forests, as is evident from the folk tales, myths, lores, idols, totem-spirit worship. In a period when land was abundant such traditions were supportive of their economic system as well as ecologically viable and sustainable too in the long-run. The autochthons, therefore, were not conscious about wastage or destruction of resources. In such a socio-economic set up the Forest Act (1885) and the subsequent royal promulgations came as a rude shock to them. The regions which they considered to be their exclusive domains got vested with the state and their movements were restricted too. They resented the 'Jangal Abadi' scheme as they viewed such a royal stricture to be dictated by ubiquitous power centres. The rights of the dwellers, though small were ignored forcing them to yield to the formal legal sanctions of the state. This hit the tribal economy quite hard where a monetised economy was introduced by this time and the need for 'using' minor products of the forest or continuing shifting cultivation was increasingly felt. Initially the government did not entertain such undiluted rights of the inhabitants, neither did it bother to change the traditional practices. However, fear of revolts and non-cooperation made them reach a tacit agreement with the local inhabitants so as not to disturb the tranquility by unqualified assertion of state rights. A policy of reserving land for the autochthons was undertaken by the king during 1930 and 1940s where in Tribal Reserves were created.[16]

Around the same time, communal disturbances, economic depression and social insecurity forced a large number of plainlander Bengalis to seek refuge in Tripura. The state administration had to provide them with land for settling as well as earning a livelihood. In 1940, 530.05 sq. miles of forest area was opened up which was increased to 630 sq. miles and still later to 1160.05 sq. miles for the said purpose.[17] As these plainlanders settled, they needed land for pasture, roads, homesteads, wastes, leaving aside land for cultivation purposes. So occassional inroads into the forests were but natural, whenever the need arose, thus bringing more and more forest land under cultivation. Despite framing of policies with the view of conservation of forests, norms were violated at random by the policy makers themselves wherever required. Such transgressions marginalised the forest dwellers further until certain forest regions were thrown open to them to collect forest products freely in 1943 (1353 T.E.). A few years later (1946-47) the Forest and Customs department of the state issued notification (10-8-1356 T.E.)[18] according to which shifting cultivators were allowed to practice shifting where they had raised crops five years back. Continuing such appeasement policy another new proposal was made, 323 sq. miles and 111 sq. miles of forest land was released in Dharmanagar-Kailashahar region where both shifting and collecting of forest products were allowed. The beneficiaries were the five hill tribes namely, Puran Tripura, Noatia, Jamaita, Reang and Halam mostly. Soon all these land proved insufficient with the continuously rising pressure of population, and so another 49,92,000 hectares of land was reserved for the autocthons, another evidence of violation of self-imposed regulations by the policy-makers.

Meanwhile, the part played by forests in preserving the natural equilibrium and moulding the state's economic foundation came to be better understood and the National Forest Policy of independent India was formulated in 1952.[19] This emphasised' a system of balanced and complementary land use under which each type of land is allotted to that form of the use under which it would produce most and land deteriorate the least. Such a policy would, it seemed, act as solution for the ongoing land problems. The policy also stressed upon the fact that it was the responsibility of the state to promote and protect forests. It enjoined that 60 per cent of mountain tracts liable to erosion and 20 per cent of plains land in the state should be maintained under forests for a proper ecological balance. Consequently, during the pre-plan period a total of 28.36 hectares were afforested which increased in the plan period. In the First Plan period 496.90 hectares and in the second 1497.56 hectares were afforested, quite a notable achievement in a tiny state like ours within a time span of less than a decade (1951-60). Along with this the boundaries were demarcated, land reclaimed, degraded forests rehabilitated and soil conservation programme undertaken too, and the forest area comprised of 8956 sq. km. Not all the hilly slopes could be afforested with tree forests, hence those with more than 20° gradient were covered with either bamboo bushes or low vegetal growth.

Forest Conservation or Settlement Programmes - A Dilemma

As the ownership of the state over forests got consolidated and became formal the special relationship which the forest dweller enjoyed in the yester-years was no more appreciated. Their rights were no more recognised, rather viewed as 'burdensone', a constraint on the path of development as well as harmful to the ecological balance. At the same time it should be kept in mind that the tribal society had undergone a sea-change in all this time, their dependence on forests are no more in totality but only marginal, to supplement their incomes. Yet, the stringent policy of the government of keeping away the tribals from the forests during the ongoing transitionary phase makes it difficult for them to cope. They are not in a position to accept the advanced technology of simple forestry operations too, and have to face a struggle for survival.

Protecting the almost destroyed forests by formulating policies seemed quite fine an attitude until the raging problem of over-population in the state was noticed once again. As displaced people poured in from neighbouring East Pakistan, they were settled on the fringes of the forests or even inside them leading to the opening up of the forests once again. Thus, population pressure led to conversion of forest areas into arable land, colonies, roads, projects etc. Those who enjoyed shifting rights in these forests had to recede to interiors and continued shifting having no other option to earn a living. A process of simultaneous deforestation and afforestation started. Forests which were deemed 'reserved' previously were de-reserved as is seen here-four regional forest areas Ramkrishnapur, Chandrapur, Garjee and Kachigang had acquired reserved status in 1936-39 as per royal promulgation.[20] But in 1960 they were declared not properly constituted and given simple 'protected' status which meant unencumbered land with good forest cover. or to put it simply the entire area of 2532.98 sq. km was actually released to accommodate the displaced people who were mostly of the cultivating stock. Therefore, it was found that from an enormous figure of 8956 sq. km of forest land in 1951-52 the cover area slid down to 5920.30 sq. km in 1981-82 due to such reckless opening up. This perhaps proves the fact that even after the policies and enactments the forest did not get much attention during the initial planning. Moreover circumstances peculiar to Tripura forces the laws and regulations governing the forests noticeably the Reserve and Protected forests to be sparsely applied.

It is evident from Table 6.2 a sample case of a decade which shows forest land is often required to be opened up for various need of the human civilisation.

Table 6.2: Opening Up of Land From Forests

Year	*River Valley projects*	*Agricultural purpose*	*Road cosnt.*	*Industries*	*Misc.*	*Total*
1973-74	552.40	79.12	0.10	–	5318.53	5950.15
1974-75	–	6.07	–	–	5458.82	5464.89
1975-76	–	489.68	–	–	8489.36	8979.04
1976-77	–	80.93	4.54	–	2054.53	2140.00
1977-78	–	282.29	–	–	1433.05	1715.30
1978-79	–	48.40	–	12.14	828.44	888.98
1979-80	–	3325.70	1.34	–	626.08	3953.12
1980-81	–	1044.16	2.21	10.12	1117.83	2364.42
1981-82	–	–	–	–	–	471.32

Source: Annual Reports of the Department of Forestry (various issues), Government of Tripura, Agartala.

The most important being livelihood, which is agriculture in this case. We can see that within a span of 8 or 9 years 5356.35 ('000 ha.) were released for agricultural purpose with governmental knowledge, while land was released for miscellaneous use most. This is a clear indication of the decline in forest land and accounts for the fact that land under forests had reduced to quite an extent which led to giving a second thought on forest conservation (1952). It needs to be pointed out here that during these years hectic settlement programmes were on and obviously under miscellaneous head land must have been given away for rehabilitation too.

It should be borne in mind that in an agrarian economy where the bulk of population depend upon land allotting land for forest cover exclusively is impossible. It is impossible to find a single reserved or protected forest where forest products are not used or where shifting is not being done. Need to step up agricultural production and to satiate land hunger of the masses have led to forest denudation and illegal cultivation over large areas leaving aside granting of grazing facilities which are absolute necessities. Strictly following the norms for conservation of forests would naturally restrict its availability. Catering to such demands, the state Forest Department has relaxed their regulations and forests of all status, except plantation land, would be opened for regulated grazing according to discretionary powers of the Divisional Forest Officer. The collection of fuel wood has been a matter of conflict among the users as prices soared to skylimits, certain concessions are granted here too. Bona fide householders inhabiting the fringe villages near the Reserved Forests can collect their requirement of firewood free of cost. Such concessions often resulted in misuse of privileges as indiscriminate fuel collection combined with timber extraction meant free access to the forest land leading to forest degradation ultimately. For development purposes building of irrigational projects involve submergence of large regions mostly falling in forest ranges (e.g. Dumbur Project). Population pressure and the absolute necessities of mankind forced the state government to rescind the 1952 notification of the Forest Policy.[21] The Forest (conservation) Act, 1980 was passed meanwhile which imposed further restrictions on de-reservation of forests for non-forests purposes, or to put in other words, clearing of a forest land for any purpose other than re-afforestation was prohibited. A new scheme of man-made plantations and afforestation was embarked upon, which soon extended the area under forest cover far and wide. Reports of the department also portray that despite loss of 37 sq. km forest cover for several reasons including shifting cultivation till 1993 there has been a substantial gain of 3 sq. km of forest cover due to regeneration in the abandoned shifting cultivation areas, as seen in Table 6.1.

Perhaps state ownership of this natural resource has been the chief inspiration behind the extension of forest cover inspite of the various bottlenecks arising from population pressure. Tripura's topography is such that dense forests are rare but vegetative growth, bushes cover low hills which though does not add to state revenue can help to balance the ecological environment of the state to some extent. The state being economically backward still, not much incentive can be provided to its farmers so that they change over to other advanced technologies yielding high productivity in terms of crops. Though concessions have been granted in the Forest Conservation Acts, these could not check the infiltrations or even regular encroachments.[22] The hill ranges, swamps, hillocks hinder extensive forest cover, the reserved forest are of various sizes–generally measuring 50 sq. km in extent with the largest being 653 sq. km and the smallest only 0.233 sq. km and these too are mostly scattered widely.[23] However, one cannot forget the rising demand for food crops in this tiny state where the population size is bursting at the seams. Therefore, extensive agricultural activity has to be undertaken which involves acres of land for agricultural need solely. A sustainable development means a rational use of the dwindling natural resources and in so doing the dilemma need to be solved first.

Field Study

A field study regarding the shrinking of cultivable land to extend forests was undertaken in a reserved forest and a privately owned plantation. Certain facts regarding the dilemma came to be noticed here.

In the reserved forest namely Trishna Wild Life Sanctuary measuring 194.708 sq. km. There were quite a few species of trees (Sal, Garjan, etc.), but it was quite astonishing to find that acres and acres of such reserved area did not have any strict boundary. On surveying the forest it was also found that 15-20 families (Hindus, Muslims, indigenous as well as settlers) occupied some land, had built houses even. They were landless and did not pursue cultivation in the forest land, only collected forest produces. Those inhabiting the fringes of the forest made regular visits to the forest in search of fuelwood, dry-leaves. However, there was clear evidence of denudation as grazing went on unheeded. The government have to keep employees with an eye on such encroachments which is an additional burden. The tribals do have certain privileges but they often misuse these and as a consequence the Reserved Forests remain so in name while the border areas keep receding.

However, a visit to a privately owned Rubber plantation in Kanchanmala, in the Sekerkote region gave a completely different picture. The plantation was spread over both hillocks and plains land and the workers were almost all tribals. They were given a little training regarding the collection and processing of rubber sap and seemed quite proficient in their jobs. There were both male and female workers who seemed to enjoy their work. The plantation had a nursery as well as produced rubber sheets for using in manufacturing finished products. Speaking to the workers regarding their present job one gets the air of satisfaction. Their fathers and fore-fathers had been shifting cultivators but with shrinking of land availability they had to turn to other jobs. With the opening of such plantations in their vicinity they felt relieved and could easily earn their livelihood. When queried whether they themselves would like to own such plantations as their proficiency could well lead to that (provided with capital) they opened that settled cultivation would be their preference. A synthesis of rehabilitation and ecology seemed to be the outcome of such plantations–it was not rubber trees alone which enhanced the greenery but on the periphery of the estate several other trees acted as wind-breakers which were the natural abodes of birds too. Other than maintaining ecological balance such privately owned plantations offered jobs to many, arrested coverted shifting cultivation and if need arose could grow tree and crops together or even grow grass to rejuvenate the lost soil nutrients. Perhaps this will be an ideal way to bring a synthesis between the available land and increasing population–a balance may be struck between the user and the land as well as getting a profit out of the same.

Settlement Programme and Forestry - A Synthesis

The foregoing analysis shows that the abundant lush green forests in the past remained unattended until the pressure of revenue was felt by the royal administration and the former's revenue generating capacity realised. But as the British Raj enhanced revenue collection and the royal expenditure too increased substantially, the need for preserving forests for the future generations was realised. Situation underwent a sea-change when the state administration began to assess the resources in modern economic terms in a broader perspective. The state began to assert its rights while the forest dweller believed that the forests were 'used' by him and rightly belonged to him. But the latter had to yield to the legal sanctions of the state and thus started a continuous process of ousting the dwellers. Already, the earlier communal ownership concept had given way to individual ownership system and the same happened in the state owned forests too. The emergence of local 'landlords' in the guise of contractors of forests threatened the security of the forest dwellers even. It came to be noticed that in the protected forests land under constructions are transferred from forests to 'khas khatians' wherein no prior clearance need to be obtained from the Forest Department. This show that the state has the ownership while the individual has the right to farm the land. There remains a divergence between the legal aspect and peoples' own frame of thinking; between the users and the conservationists, between development and forestry. All these factors ultimately make two elements primodal i.e. the forests and the population, wherein a synthesis has to be struck so as neither is ignored in the process of pursuance of the other.

The plantation scheme gained a stronghold in the state agrarian structure since the 1960s. The actual aim was to utilize degraded large hilly tracts as a means of weaning away the nomadic shifting cultivators to a settled life, and by doing so reduce the pressure of population on land as well as restore the destruction already committed. The Tripura Rubber Plantation Corporation was conceived in 1963 to rehabilitate 800 landless shifting cultivators on 1200 ha of rubber plantations.[24] This was further increased to 4646.82 ha in 1982-83, which was aimed to serve the purpose of creating a synthesis between the population pressure and land crisis. They provided jobs to several forest dependents and landless labourers. Interestingly, data given in Table 6.3 show that a majority of the workers are tribals giving away the fact that these autochthons were mostly forest dwellers and hence were most affected because of forest conservation acts and policies. Table 6.3 clearly reveals a bare fact that there is trend of increasing dependence of agricultural labourers on plantations in a changed environment. Rubber growing has found a congenial climate and soil in Tripura and those regions which had turned barren due to over-use and shifting were subjected to extensive Rubber Plantation Corporation, where in joined the landless cultivators leaving agriculture for the privileged class only. Another feather added to the states plantation policy was the Tripura Rehabilitation Plantation Corporation which was set up in 1983-84 for rehabilitation of 800 landless jhumia families.[25] Each family was allotted 2.5 ha of government khas land (mostly uplands) of which 1.5 ha was put under rubber cultivation and remaining 1 ha for housing, horticulture, silviculture, apiculture, pisciculture etc. Their published reports show that they have been able to achieve their targets as they had estimated, certain private individuals have also taken up such plantations and the climate congeniality as well as soil bed have enabled them to reap good profits within a decades time. They have engaged quite a large number of local workers who were landless previously but working in the plantations they have acquired certain land for building houses and even for pursuing settled cultivation as well as becoming semi-skilled and skilled labourers in their own field. If plantation projects are taken up seriously and the plant species are diversed too with an eye to its congeniality perhaps the land crisis may be arrested to a certain extent. The continuously rising population, of which unskilled labourers form quite a huge percentage may be roped in as labourers with minimal training. Of course there have been complaints regarding the working pattern of the autochthons. They find that these new forest plantations have no utility for them rather hinder their easy method of shifting inside the forest promises. Though they do work in the plantations they seem to be content with being able to earn two square meals a day or spend a handful during festivities, but they deter from working regularly as is revealed from the field studies. These plantation trees do not bear fruits which affects the tribal economy quite adversely. Their reluctance in coming out of their self-styled cocoons make them unemployable at times, since the present operations cater to a wider need. The tribal workers have increased to a great extent undoubtedly (Table 6.3) but one cannot overlook the fact that the non-tribal workers are increasing at an equally fast rate. The synthesis, which is being worked at through plantation schemes is to be instilled amongst the forest dwellers (both tribals and non-tribals) because, it will not be possible to contain the pressure on the forest unless the dwellers get the feeling of oneness with these projects. A way out to solve the dilemma is that not only should the government indulge in economic upliftment but have enough infrastructure to back it up too.

Table 6.3: Number of Workers (Tribal and Non-tribal) in Rubber Plantations

Year	*Total workers*	*Non-tribal*	*Tribal*
1976-77	92,555	NA	NA
1977-78	1,64,627	73,132	91,495
1978-79	2,27,213	1,03,886	1,23,327
1979-80	3,84,831	1,96,658	1,88,175
1980-81	4,42,889	2,14,995	2,27,894
1981-82	4,74,868	2,00,890	2,73,978

Source: Tripura Forest Development and Plantation Corporation, A brief Report (1984).

In following such a kind of synthesis programme, the dichotomous nature of the state population comes in the way. The local inhabitants consider forests to be their domain and the presence of a large number of immigrant labourer (non-tribal labourer) may force an untoward situation of communal misgivings. A total reservation of forests could not be done in Tripura in the face of pressure of population on land. But whatever reserve was created did oust some of the users, which is not at all welcome. Acts and policies of the forest department should not disrupt the base of the indigeneous economy. Because if this cannot be avoided those who are essentially forest dependent in some stage or another will be forced into the substratum of the new socio-economic structure already established by the immigrant population.

It is quite clear that the existing forests are still inadequate to meet the needs of the people and it is imperative to both strengthen and expand their extent. The existing reserves could not be opened up completely but it has not been able to thwart encroachments too. A middle way may be sought, those who depend upon this resource be given full authority to maintain them too, since they have the potentiality of the greatest preserver. The path of development is quite slow in this state due to lack of adequate financial support which could have been helpful. Planting of trees utilising the open land within forests for growing crops, collecting the minor forest produce (which need not be treated as state income but as an element to provide economic relief to the shifting cultivators at least during the process of transition particularly), can give a means of livelihood to the agriculture dependent population who have lost their rights over the forests owing to conservation Act. The 'user' is generally looked upon as an essential encroacher and not taken into confidence and entrusted certain duties too. As such, there is a constant threat of the contractors who direct the produces of the forests for their own interests notwithstanding the prohibitions even. In such a situation a synthesis between the forest preservation and population explosion may be attempted by changing the system of ownership, the forest user may be given the ownership of the natural wealth. Those extensive barren or waste land may be given to the communities to restore ecological balance. The users, who in most cases are local inhabitants have an inborn love for the wilderness and there is no reason why they shall not be interested in growing trees. Granting of properietory rights, initially in a limited manner, by training them in utilising the forest wealth, making them aware of the economic species of trees, they can be easily brought into the programme where forest conservation and rehabilitation of the populace will go hand-in-hand, neither being sacrificed.

A third way of creating a synthesis is perhaps by having a forest near agricultural fields. A good forest cover implies that all the natural forces like water, air, soil and animal life interact to produce the maximum amount of leafy matter, timber, flora and fauna. The forest helps to trap rain water and channelise it into under-water streams, acts as a barrier against wind erosion (wind breakers) thus help to retain the fertile top soil as well as acts as a barrier against pests too (being full of birds, animals, insects, reptiles etc.). But planned forests, under private ownership often end up in planting of the same kind of species where the roots are of the same depth, trees planted at equal distance leaving unoccupied space underground which can move down hill in a heavy shower. Single species forests, therefore, tends to be sterile and poor above ground and loose and ineffectual underground. Again native species are more congenial (teak) as it attracts birds and insects where as exotic species (eucalyptus) do not help much and continue to be simply a plantation, even prone to pests at times. Therefore, an attempt at planting natural forests of different species need to be harped alongside the fields.

Development activities in terms of jobs, cash are more tangible and lucrative to the indigeneous people of the state but at the same time forests need to be conserved too. A compromise to stipulate forest land for using as orchards or tree farming may be useful. Wood, timber demands too are growing fast. For this every village or town should have their own plantations which will be possible with cooperation, discipline and organisation. Such cooperation is not difficult to find among the tribal people, what is needed is proper infra-structure backing, those who have to depend upon forests may be given employment or a share of revenues for the job done. Undoubtedly, their labour will not be skilled with no benefit for the revenue reapers but at least they will be trained to look after the forests and simply not waste the natural resources.

An Integrated Plan

In the previous century, more than half of the total land was uncultivated. Later, these uncultivated land near the villages became cultivable plots, grazing lands and those farther away turned to be forests under state administration. Gradually several rights of the people using this land were extinguished but growing population pressure did not allow these forests to remain 'uninhabited wilderness' and it became a source of livelihood for scores who lived nearby. The Conservation Acts and policies were evolved, which at present seems to be a evolution in the historical perspective. In a small state like Tripura, where there is a massive population pressure on land, the ensuing problems magnified within a few decades time. Environment was given more importance, alongwith forest wealth's commercial value and thus started the process of reserving and protecting forest. Land being rich with soil nutrients, excellent climate, growing forests is no problem, but the forests are under tremendous biotic interference which threatens the ecological balance.

Hence, a shift in emphasis regarding conservation is needed. Afforestation is undoubtedly most needed at present but not at the cost of cultivable land. Raising of seedlings, collection, processing, transport of forest products can engage a large number of people and these conservation can go hand-in-hand with rehabilitation. The administration has given in to certain demands of the people (collecting fire-wood, grazing facilities, etc.) making the purview of acts broader by these concessions and privileges. Such requirements can be met with community forest grown by the people themselves. So long fodder and fuel wood come free farmers, even if wooed into planting of community forests will opt for revenue generating trees only.

Forests are an essential feature of the land question of a state yet it becomes a problem where the population requires space for rehabilitation as well as development. The dilemma will remain, until conservationist plans and policies can be made more sensitive. A shift away from state monopoly is an essential precondition for fulfilling both ecological balance and livelihood security. If commercial timber production can be diverted to private farm lands, it has been seen that much less land area (almost one-fourth) will be required as compared to the area of natural forests requirements. At the same time active involvement of people will help remove inherent conflicts. Those inhabiting forests have no doubt played a part in denudation of forest indiscriminately, but to put the blame on them solely is only a convenient diversion. Emphasis has to be put on afforestation by planting diverse species on available area and supplement them with grasses, shrubs and bushes, which could restore the old harmony between need of the market and peoples' exploitation. The basic issue which often gets tangled in Tripura's context is the dichotomous character of its demographic structure, where the autochthons find it hard to adapt while the settlers are more keen in doing so. None can be left out in the process of economic development, hence the need for an integral approach. Initially, the requirement of fulfilling the basic needs should be catered in the specific socio-economic context. Plan for simultaneous development, with social integration, people's active participation will help in solving the land availability crisis in the state.

NOTES AND REFERENCES

1. Ramaiah, P.: *Tribal Economy of India,* Light and Life Publishers, New Delhi (1981), p. 81.
2. Agarwal, A.K.: *North-Eastern Economy, Problems and Prospects,* Mittal Publications (1988), p. 165.
3. National Remote Sensing Agency: *A Report, Satellite Remote Sensing Survey of Natural Resources of Tripura,* (1985), Vol. 6, No. 3 and 4 pp. 517-528.
4. This is from an estimate of the Forest Survey of India (Ministry of Environment and Forest) Kaulagarh Road, Dehra Dun, 1993. Based on the 1980-81 to 1984-85 forest cover it was seen that the loss in forest cover due to shifting and other reasons amounted to 983 sq. km in the North-eastern region and gain due to regeneration in abandoned shifting cultivation areas and other reasons was 348 sq. km This negated the gain eventually totalling 65 sq. km. of loss in forest cover. At the same time Tripura lost 10 sq. km due to shifting cultivation and gained 15 sq. km due to regeneration while lost 27 sq. km gained 25 sq. km due to other reasons totalling a gain of 3 sq. km in forest cover.
5. Chakraborty, M. Ganguly, A and Das Gupta, M. (ed.): *Development of Forestry in Tripura* (A Compilation of Seminar papers of NEICSSR) Omsons (1989) - editorial note.
6. *Annual Administrative Report of the Political Agents* (1874-75) Prepared by Capt. W.L. Samuells (WBSA).
7. *Ibid.*
8. Calcutta Gazette (1865): Act VII, Section-2 (WBSA).
9. *Ibid.*, Issue dtd. - Feb., 1871.
10. *Ibid.*, Act I, 1296 T.E. (1886), WBSA.
11. *Ibid.*, Act I, 1297 T.E. (1887), WBSA.
12. Tripura Code, Part III, 1909 T.E. (1899) - 'Jarip-o'–Bandobast Sambandhiya Niyamabali.
13. Hunter, W.W.: *A statistical Account of Bengal,* Vol. VI, Report of an Officer of Topcgraphical Survey, London (1876), p. 505.

14. *Ibid* (Permit system of 1903 was issued by the king of Tripura).

15. Several area were already being brought under Reserved Forest or unclassified protected forest areas by the Bankhand Reserve Memos of the king in Langai, Machmara, Unakoti region where shifting was being prohibited. But due to protests from the essentially shifting cultivators such land had to be opened up hence the need for reserving 'trees' rather than 'areas' was felt No. 223, dt. 16-6-1353 T.E., Tripura State Gazette Sankalan.

16. The tribal reserve orders of 1931 and 1943 specified the area of about 110 sq. miles and 1950 sq. miles respectively for settling the tribals (in Tripura state Gazette Sankalan, Directorate of Education, Govt. of Tripura, No. 223 dt. 16-6-1353 T.E.)

17. Tripura State consolidated Administrative Report (1347-50 T.E./1931-40 A.D., 1350-52 T.E.) 1940-43 A.D. and 1353-56 T.E./1943-46 A.D. (TSA).

18. Tripura State Gazette Sankalan: *op.cit.*

19. National Forest Policy, 1952: A Report, Ministry of Food and Agriculture, Government of India.

20. Maharaja Bir Bikram Kishore Manikya's notification No. 378/3/87 dt. 6-1-1349 T.E.

    (i) No. 7853 B/3-83 dtd. 21-1-48 T.E (Chandrapur)

    (ii) No. 35 B/3-85 dt. 6-1-49 T.E. (Kachigang).

    (iii) No. 4160 B/3-65 dt. 10-10-46 T.E. (Garjee).

    (In Tripura State Gazette Sankalan *op.cit.*).

21. Department of Forestry No. F. 2-2/5/For-80/CD dt. 9-8-82.

22. As the communal ownerships gave way to private ownership of land a middle-order emerged, that of contractors/landlords. In this case it comes to be noticed that while state has the ownership, the individual has the prerogative to use the land. Land under continuous constructive possession in the protected forests are often transferred from forests to khas khatian under which no prior clearance need to be obtained from the Forest Department. These middle-men often look out for unclaimed land from where trees can be extracted free of cost. Acres and acres of land have been allotted indiscreetly flouting all rules and regulations.

    (Excerpts from field studies conducted).

23. Reports of the Department of Forestry, Govt. of Tripura, Agartala.

24. Tripura Rubber Plantation Corporation; A Report, Deptt. of Forestry, Govt. of Tripura.

25. Tripura Forest Development and Plantation Corporation: A Report, (1980) Deptt. of Forestry, Govt. of Tripura.

7

Impact of the formation of the Tripura Tribal areas Autonomous District Council

The very term autonomy means one's own right to self-governing, mainly relating to a community. Autonomy often follows independence—they are complementary concepts in a norm, however, it should not be related to political area alone, 'but should be expanded in every sphere of life in order to enjoy the freedom to explore life without sacrificing essentials of a meaningful tradition'.[1] It may be said in simple terms that when a community can take its own decisions regarding administration, without being guided or pressurized by any other community the former may be termed as Autonomous. A kind of self rule which generally works within a constitutional framework. Autonomy movement is a consequence of any apprehensive feeling of a minor community towards a bigger and more powerful one where in the smaller community tries to free itself from the shackles of exploitation. Such movements can be witnessed in regions where there are a lot of diversities viz., racial, ethnic, cultural, lingual etc., but the inhabitants reside in a definite geographical territory.

Autonomous states have been heard of in India as long back as 6th century B.C., but a movement for the same is only a recent phenomenon. As the British tightened their grip over India, the leaders of the nation agitated for self rule. When granted

Independence the various groups pressed for a share in the power and direction. Unlike the rest of India, the tribal states in the North-east region were not linguistic states but ethnic states, although the ethnicity of some of them is not uniform. Most of the population inhabiting this sector are not indigenous people but migratory waves from neighbouring countries and states dating back to time immemorial. The British annexed Assam to their Indian realm in 1826, and the states bordering the North-east region came under their control as its natural resources (timber, ivory, muga etc.) allured the East India Company to develop trade interest in this region. A new form of administration the Non-Regulated System was evolved aiming at protecting and maintaining the exclusive culture, customs and the traditions of the indigenous people, setting up an administration suited to their genius and consistent with their ethos and aspiration. By several laws, regulations, orders the tribal society was ensured protection of its identity and culture, preserved their socio-religions norms, values in its pristine form. Yet, the British policy aimed at confining the autochthons in their hinterland as they tightened their grip over the mainland of India. They viewed the tribal areas as colonies of exploitation to serve their own economic and political objective. The very term 'Tribal Areas' was first introduced by the Government of India Act, 1935.[2] Such areas were given a special status by the Foreign Jurisdiction Act of 1890. But when revised in 1937 the Governor of Assam could discharge his functions as the agent of Governor General of India in Council. A zone of political control was created which gradually shifted to the interior hills. Such behaviour meted out by the British instilled a sense of fear in the tribal mind and somehow, the aborigines never participated in mainstream national politics during British hegemony. They seemed to be content with their own customs, economy and other aspects, absolutely unaware of any nationalistic feelings. They thought that any sort of political or socio-economic merger would rob them of their traditional self-governance and swamp their economy through entry of the outsiders. Even the political parties of the country did not try to integrate them. Subsequently the transfer of power led to the abolition of 'special' status of tribal areas. Any reference to these were altogether dropped in the Provisional Constitution of 15 Aug., 1947. Here, mention must be made of the provincial jurisdiction ordinance which called on the tribal areas on the frontiers of Assam as Assam Tribal Areas.[3] The policy of total segregation of the autochthons was rejected and steps were being taken to bring them within the mainstream. Perhaps, the economic prospects of this North-Eastern Sector, oil fields, forests, had attracted the attention of the government. A policy of "opening up" of the tribal areas was undertaken, and this intrusion by dominant patrons broke down the cohesive bond in the traditional society. They realised that to live in the new world they had to live a new life. The tribal penchant for managing their own affairs according to their customs and their inherent fear of being dominated by an alien people led to the break out of armed rebellions from time to time. Coming in contact with these settled people, the creamy layer among the tribals got enlightened and educated and gradually replaced the traditional leadership structures even. Voices were raised and demands made for participation in the policy making process too so that they stood on an equal footing with the rest of the fellow citizens. It is quite obvious that this discovery of identity was facilitated by the presence of newcomers from the plains. This ambition of the aborigenes to take part in decision making, apprehending the loss of their land, language, and culture made them demand Autonomy, and the four tribal districts of Assam – the Naga Hills, the Lushai Hills, the Khasi and Jaintia Hills and the Garo Hills attained autonomous Status of Nagaland (1963), Mizoram (1986) and Meghalaya (1972).

This chapter will deal with the tribal autonomy movements in Tripura with a historical account of the autonomy movements in the state which led to the formation of Tribal Areas Autonomous District Council and analyse its impact on the land question.

Tribal Movements in Tripura (19th and 20th century) towards Autonomy'

Tribal movements for autonomy in Tripura was the same as that of the North-east, however due to the state's peculiar character it was not identical. The monarchy being basically of the tribal ethnicity the hill tribes were never completely ignored.

The chief of the Tripuri tribe, subjugated the other tribes (viz., Kuki, Halam, Reang, Jamatia) and established his authority, though following the traditional tribal mode of administration and even granted enough freedom to the other tribes. With the ushering of modern administration during the late 19th century, however, the king found that he could not manage with tribal chiefs solely. Hence the assistance of the neighbouring East Pakistani nationals was sought and they were inducted in the state's administration as is evident in Mackenzie's narrative–"The persons composing the government or possessing the chief management of the country are the Hindus—called Senaputty and Burroahs—"[4] Some autochthons did remain, acting as links between the government and the inferior vassals. The indigenous masses who thought the king to be one of them now grew distant and looked upon him with awe and respect which did not imply whole hearted loyalty though. Thus started a gradual estrangement of relations between the king and his actual subjects which widened as the king started pressurizing them to pay increased taxes. The peasants often felt exasperated by such exactions and had to even forfeit their lands, resulting in occassional revolts against the state. The hill people did not have to pay any rent for carrying on shifting cultivation, except a house tax, the Kukis rendered military services and so they were exempted sometimes from paying taxes which was fixed as Rs. 5-4 annas and the Tripuris used to pay less, (Rs. 3-8 annas) since they worked for the royal family. While the Noatias and Reangs had to pay Rs. 10/ per family; Halams paid Rs. 2/- 10 annas or less per family and the Jamatiyas paid Rs. 3-8 annas per family, along with the rent collectors' fee which amounted to 50 per cent of the original tax even.[5] Such inequality in tax structure bore seeds of injustice and discontent brewed more so because no steps were taken for poverty alleviation, checking famines, pestilence or providing security against attacks. Such being the circumstances, alienation in every aspect of life was obvious. Samser Gazi's revolt in the 18th century had been one such revolt since even the Mog community supported him.[6] A century later, in 1850, the king's own ethinc group rose in arms against such oppressions[7] even the Kuki raids of 1860s were not pure plunders, rather, a wider movement in which certain sections of dissatisfied subjects also participated; the Reangs who were facing much hardships due to crop failures had to borrow from the money-lenders and their debts piled up, they too joined the Kukis,[8] the Jamatiya revolt of 1863 had an inkling of deep rooted crisis in the economic life of the people. The District Magistrate of Tripura opined that it was because of the king's exactions of rent collections.[9]

Analysing these revolts' character, in the historical perspective, a point comes clear, the movements were aggravated as the kings continued to invite and pamper the Bengali immigrants by giving away chunks of plain land either at absurdly low rents or even rent-free. Abnormal population increase and the shrinkage of cultivable land provoked mass movements in the 19th century. The autochthons got subjugated to the immigrants. They could not practice their traditional shifting cultivation and had to relinquish their hold on irrigable land and move to the interiors where hills became their domains and their method of cultivation did not seem conducive anymore.[10] On the social plane also they felt a change in attitude of the kings who seemed more inclined to adopt the immigrants' religion too, thus a deep distrust between the two communities residing within the same territory evolved a conflict between the autochthons and immigrants.

In order to make up for the worsening relations the royal family decided to act fast. The rulers undertook extensive tea cultivation which became a major source of revenue and also offered employment to many.[11] Yet the essential autochthon could not find a place for himself here being unable to adapt to the ongoing change and shied away. Immigration continued unabated, the then king, Maharaja Bir Bikram Kishore Manikya tried his level best to wean the tribals to plough.[12] He even proposed a new constitution (15 April, 1939)[13] and introduced the Gramamandali Act in 1940,[14] along with spreading of literacy among the youth, a job which the Christian missionaries had undertaken under the garb of emancipating the simple indigeneous people from socio-economic bondage. Education made the youths conscious, and they began to analyse their place in society.[15] The people's complaints against the oppressive nature of administration found vent in the Reang revolt of 1941-42 as the seething troubles of the tribes burst. The political organisations which were gaining hold since late 1930s (The Tripura Rajya Gana Parishad, Tripura Jana Mangal Samity) aimed at socio-economic upliftment of the people and soon got grouped into two wings of the prevalent political ideologies. A new horizon for tribal movements in Tripura opened. A few communal organisations secured a stronghold in the state, viz., Tripura Sangha, Anjuman Islamia, Bengali Hindu Sanmelan etc.[16] The Tripura Rajya Praja Mandal was formed for self determination of the Tripuris but not entirely by the Tripuris.[17] Education caused the youth to spread the 'mantra' of ethnic identity which caused great furore in the political arena of the state since the population of the state did not belong to the same racial stock. The immigrants too were aware of the raging political activities, coming from a hot-bed of revolutionary activities (East Pakistan) and political ideas seeped into tribal belts. Amidst such a political turmoil, the state administration was taken over by the Indian Dominion in 1949.[18]

A light of hope percolated among the indigenous inhabitants of the state who looked forward to their own socio-economic upliftment. But by this time population pressure grew by leaps and bounds and there was a mad rush for land acquisition among the immigrants. These new comers were settled in the indigeneous peoples' secluded corners even, those lands which the inhabitants failed to put to proper use due to lack of skill or ignorance were grabbed by the plainlanders and by the end of 1957 about 3,65,000 immigrants entered Tripura which broke the last remnants of isolation.[19] Appeasing techniques of the state soon allotted cultivable land to the refugees instead of attempting to quell the problem as they explained that the tribes did not always depend upon the income of plain land[20] where as the immigrants knew proper utility of the same, fanned it more. Here is an example of such an allotment–a plot of tilla land consisting of deep forest (approximately 1 drone) was allotted to one M.G. Sardar by royal orders. He had cleared and made it fit for cultivation, but this land was taken away for refugee rehabilitation purpose by the Department of Rehabilitation.[21]

After independence and following merger, the thrust was however, on the demand for protection of the tribals' right to land. So after quite a chunk of the 'reserved' area was 'de-reserved the Tripura Rajya Upajati Gana Parishad demanded that now the remaining Reserve should be well demarcated so that it might remain protected.[22] The then Member of Parliament, Mr. D. Deb (1952) suggested to the Prime Minister, Jawaharlal Nehru–"Some area or areas of Tripura shall have to be set aside for the tribals alone and no other persons should be allowed to settle there."[23] A few years later, it was demanded that this reserve area should not be further released and the emphasis was on the introduction of regional autonomy.[24] The Scheduled Areas and Scheduled Tribes Commission was entrusted towards finding a solution to the on-going tribal discontent, under the Chairmanship of U.N. Dhebar (1960-61). The Commission recommended the setting up of Tribal Development Blocks on an experimental measure.[25] The Administration Reforms Committee suggested the setting up of a Tribal Council with well defined administrative power.[26] The influx of refugees continued thus upsetting the local economy and the Tripura Land Revenue and Land Reform Act (1960) failed to safeguard the rights of the tribals on land. The opposition party during this time stressed on declaring a schedule area in Tripura. After sometime Tribal Development Blocks were created in lieu of scheduled area but even these blocks were deserted by the tribals as those utterly failed to give any protection to them.[27] Such discriminatory practices of the government led to certain socio-psychological problems among the aborigenes of Tripura. They were aware of the massive exploitations they had to face, modern education was able to awaken them from the slumber of primitiveness, they were already introduced to modern living and aspirations soared higher, being deprived of such a life due to wide-spread poverty, they found their existence, their 'identity' threatened due to the presence of an 'alien' race the numbers of which multiplied continuously. The sense of becoming a 'minority' in one's own land without even land ownership rights threatened these tribesmen. Whatever land was given to them due to several reasons were going out of their hands, even their labour capacity was being exploited. Therefore, the sense of self-consciousness, identity crisis and alongwith it deprivation of the basic needs of life led to a transformation in their behaviour pattern too. It was reported that large number of tribals had to leave Tripura and seek refuge in the Mainama Forest area of East Pakistan or in different forest areas of Assam for their livelihood and survival. It was obvious that Tribal Development Blocks would not solve the problem.[28] The government was all along opposing the demand for the Fifth Schedule as in its view the population of Tripura had changed its character and Tripura had turned into a mixed area. However, there were places where tribal population constituted an average of 69% of the total population—North Debendranagar, Takarjala, Chamanu, Amarpur, Kulai, Kanchanpur etc. And the demand for implementing the Fifth schedule in the predominantly tribal areas of Tripura was made,[29] but to no avail. Hanumantaiya Commission Report on North-east tribal problems was presented to the Prime Minister now demand of implementing the Sixth Schedule to the tribal compact areas intensified alongwith handing over of the developmental activities too. The hitherto 'moderate' organisations (like Adibashi Samiti, Adibashi Samsad, East India Tribal Union) could not give them the necessary justice, so they united in a stronger bond as the need for autonomy gathered momentum—the Tripura Upajati Samity was formed in 1967. In this was included tribals from all walks of life irrespective of their occupations, affiliations or sects. A new elite leadership was also created along with this political organisation the moderates, the extremists and liberals all came under one banner to march towards autonomy.

The Samiti now worked in the direction of making a more concrete and emphatic demand of land language and local autonomy. These were the chief bones of contention all through. However, amongst these, the 'land' question seemed to be the primary demand of the tribals as it was their only source of sustenance. The tribal economy underwent vast changes as their traditional systems were rendered unprofitable with the introduction of modern techniques. Though some did readily acceede to these new methods they were not properly groomed and hence could not be at par with their non-tribal counterparts in crop production. Also jhum land had dwindled, even the yield had shrunk, the land which they had tilled once had no mention in the official records of the state even so falling back on this was simply impossible. Whenever any new settlement of land was introduced the autochthons hardly found their lands registered accordingly.[30] One may blame illiteracy to be a cause of such a condition, but the dwellers put the blame on the apathetic treatment of the government. Cry for restoration of tribal land went unheeded as the tribal reserves were abrogated by an ordinance on 28 Feb., 1974 giving the reason that land was required to accommodate the displaced people after the Bangladesh liberation war (1971). Even at this juncture shifting, which was permitted to be continued by the jhumiyas (by the Dhebar Commission) was prohibited through a government notification. A third amendment was made in the Tripura Land Revenue and Land Reform Act, 1960, to safeguard the tribals. The outcome of which was perhaps earning of a 'tribal schedule area' (1975). But the demand of autonomy did not get a priority and some of them resorted to terror tactics even, which reached its climax on June 6, 1980 when hundreds of non-tribals were brutally massacred at Mandai in a day. It was clear that the pent up emotions of deprivations found an outlet through such a contemptuous bloody riot, which violently jolted out the state from its stupor.

The state government now reacted hastily and passed the Autonomous District Council Act of 1979,[32] providing practically all the safeguards and powers to the tribal communities under the constitutional provision of Schedule VII, on 18 Jan., 1982. To introduce internal autonomy in tribal compact areas and thereby protect the socio-economic-cultural interests of the tribals the long awaited Tripura Tribal Autonomous District Council Act, 1985 was passed by the Parliament under the Sixth schedule of the constitution replacing the thirty-nine months old Autonomous District Council under Seventh Schedule.

The formation of the Tripura Tribal Area Autonomous District Council (TTAADC) under the Sixth Schedule of the constitution of India is believed to have fulfilled the long cherished dream of the hill people of the state who clamoured for self-government (autonomy) and their socio-economic upliftment. The granting of autonomy was met with much fanfare and acceptance by the people. The main object behind the formation of an Autonomous council was to handover certain administrative and legal authority to the council, to pay special attention to all aspects of cultural, social and economic improvement of the indigenous communities of the state. The ADC was given a free hand in allotment, occupation or use of scheduled land or reserved forest area; management and control of any forest which was not reserved; regulation of shifting cultivation as well as money-lending operations.

Initially, an area of 7,132,56 sq. kms. was brought under ADC's jurisdiction where more than 60% of the population was essentially of the tribal stock.[33] 462 Moujas were included within the tribal area which accommodated 19 tribes in the same autonomous regional council, an unique feature in the North-East mainly because three other North-Eastern states of Meghalaya, Mizoram and Nagaland have separate autonomous bodies for different tribes. The TTAADC has been a landmark achievement in realising united demands of the tribals for autonomy. A long standing war against the state by a section of its inhabitants to gain their lost prerogatives seemed to come to an end. Their agitation for removing the social discriminations, working for social upliftment, availing of equal opportunities were answered by this notification. The tribals were brought to parallel standards with the non-tribals. The ADC gave priority to the socio-cultural, economic interests of the aborigenes. The hitherto withdrawn tribesmen, came out from their seclusion to get a share in their own governance. Things went on smoothly in the initial years even the Tripura National Volunteers' the militant outfit, seemed to cow down and decided to abjure violence and give up secessionist demands. The ADC mainly aimed at reforming the land problems in the state, specially restore those land which were given away to the new settlers or those which they transferred in their names using fraudulent means. The loss of their most important possession, their land, shows impoverishment and loss of self esteem and this is a recurrent complaint among the indigeneous inhabitants of the state. The establishing of the ADC is intrinsically related to the land question of the state, or precisely to the problem of infringement of the rights of the tribals of Tripura over their land.

The ADC was empowered with the rights to make things better for the indigeneous tribes. Funds were allotted for various developmental projects pertaining to the tribals only. Priority given to land reclamation schemes, forestry, soil conservation, jhumia and landless tribals' rehabilitation. The ADC has a Land Records and Settlement Department to distribute lands to the landless for construction of dwelling house as well as for cultivation purposes. Special emphasis is given on allotment of lands to the landless tribals as well as non-tribals who reside there. Their 'Achievement Report' revealed that during the initial years 35,718 landless were allotted land of which 2,678 belonged to the 'others' category.[24] The ADC has also evisaged some rehabilitation projects for the jhumias to wean them to other jobs like forestry, settled agriculture, animal husbandry, pisciculture, 4,251 jhumias & landless rehabilitated in 48 such projects. It had also set up 25 colonies under the Principal Officer (Welfare) ADC where 1,759 jhumias were settled upto March, 1993. Even the LAMPS* have come forward with funds to help the tribals, as all the LAMPS fall within the ADC areas. The ADC, on the state's behalf has implemented the State Rural Employment and National Rural Employment schemes also to help the tribals during the lean months by providing jobs. In the inaccessible hilly terrains agriculture and horticulture kits are distributed. Out of 7,132 sq. kms. of total area 2,050 sq. kms. are unclassified forests i.e. erstwhile protected forests and khas land. The ADC is able to manage the forests (not reserved) as per section 3(b) of the Sixth Schedule of the Constitution of India. The ever increasing population and their subsistence means of jhuming is fast depleting the existing vegetation and valuable

* Large Sized Multi-purpose Societies.

forests. Steps are being taken for social forestry and under taking rubber plantations, e.g. 6,011 hectares have been planted with teak and gamai plantations by the government, 100 hectares were allotted for rubber plantation and forest development while 4355.89 hectares are planted with bamboo, coffee, teak etc. by the forest department of the Autonomous District Council.[35]

The ADC, which was the outcome of a struggle made both by the tribals and the non-tribals of the state, seemed to do something worthwhile for the hitherto backward section of population. But soon it ran into rough weather. The dominant immigrant population found that three fourth of the state's land was reserved for only one-fourth of its population, and the majority of the people had to depend upon one-fourth of land area.[36] Earlier, the non-tribals who 'possessed' these lands had reclaimed them and made it suitable for ploughing but with the allotment of Autonomous District Council area they had to surrender the same. This seemed to divide the ADC on communal lines, instead of heading for a harmonious future it seemed to be marching forward to another catastrophe. Undoubtedly, the ADC was all geared up for doing the needful for those areas which had more than 60 per cent tribal population. But it is not necessary that all the tribal population of the state flocks to these areas only. Those living outside did not get the same privileges which meant invitation to arms struggle which had to be avoided at any cost. Large-scale presence of non-tribal refugees, created a crisis in the ignorant tribals' mind. But in the past the two segments of the population did live in perfect harmony without expressing any parochial sentiments of antagonism against each other. Autonomy seemed quite natural with the increasing crisis of land question which provided them with subsistence at least. The jhum cycles had got precariously shortened, also the newly settled tribals faced a crisis as more and more land was needed for rehabilitation purpose. The feeling of alienation and a crisis of their identity factor made the tribals demand for more autonomy as they resisted from becoming "marginal masses" in their own land.

The ADC was formed and initially seemed a perfect answer to the demand of the autochthons yet it is still a minor body. So the time is not ripe yet to adjudge its contribution properly. The ADC gave priority to the socio-cultural-economic interests of the aborigenes. The hitherto withdrawn tribal people eagerly came forward to get a share in their own governance. However, the land question seemed the most problematic factor. Though the TLR & LR Act of 1960 was amended in 1975 (Third Amendment) wherein was mentioned that the tribes had to sell their land only to the members of their community. It was soon found that this was only an eyewash as very few tribals were in a position to buy such land. Lands changed hands covertly though the administration dispossessed several non-tribals who had developed the land with hardwork. Thus, those who depended upon land found it improper, the way land was being taken away from the actual tiller. After the formation of the TTAADC those non-tribals who previously resided in these blocks but were dispossessed of their lands viewed it as a 'diabolical scheme sponsored by the lackeys of imperialists'.[37] They thought it to be 'a narrow, selfish interest of the ruling class of exploiters in the tribal society which have found expression through the demand for and establishment of the ADC'.[38] It may be pointed out here that the political actions of the indigenous people should not be looked upon as premordial, but as actions by their own conscience, shaped by centuries of its own polity and deprivations. They tried to act within a new concept of struggle with the aim of attaining autonomy.

The question remains, how much 'autonomy' has the TTAADC exercised since its inception? Reports and personal enquiries show that the "Symbol of hope" for the poor hill dwellers has hardly been able to make any headway in finding solutions to any basic problems of the tribals so far. The TTAADC had primarily aimed at ensuring the right of the tribals on land. Apparently it seems that the autonomy movement in Tripura is perfectly organised. Funds are allocated for various developmental projects. Undoubtedly the ADC has done some work in this direction of land reclaimation, soil conservation, forestry, jhumia resettlement etc. But many of its enterprising projects take a long time mainly because of governmental procedure. The ADC does not have any village council so it cannot show the "originality or innovativeness".[39] The matters concerning land in the tribal autonomous areas are disposed through a cumbersome and time consuming process.[40] First of all, the ADC receives applications from the tribal nomads, the landless and the marginal farmers who seek land allotment within the region. These applications are then sorted out according to mouja, tehsil and sub-divisions. The sub-divisional officer then allots the land amongst the applicants according to land rules of 1980. The state government also collects land revenue in the Autonomous District Council area and the concurrence of the ADC is required for allotment of khas land. During the initial years the CPI (M) led Left Front did not want to part with this authority of allotment of land in the ADC areas. Though with the changing over of the government, the Congress (I)–TUJS alliance wanted to hand over this privilege to the ADC empowering it to deal with the land issue effectively, however the order did not materialise.

Even the leaders do not show the initiative to deal with the inhabitants of the ADC zones. Perhaps the inaccessibility, poverty, ignorance, illiteracy, and motivation of a self sufficient economy hinders the remote areas' tribals to come forward eagerly seeking the developmental privileges. Hence, the council always is not able to encompass all the tribal population of the state and a handful get these privileges who are creating a new class in the society. The council does not have any unique method to implement the projects. It has to depend upon the state government for every thing, finances, project execution and subsequent implementations. To overcome such lacunae of parallel offices the council has established four zonal development offices,[17] sub-zone offices to carry out the programmes. It is duly assumed that the ADC officers and employees will understand the tribal's woes and problems more acutely than the state officers and a conflict between the two may be avoided thus and work could go on smoothly.

However, the revenue from the zones is still being collected by the state government. It is the usual story, budget provision and actual allocations are never the same. The deficit in the allocation hampers the ADC from taking up any development projects. There is also another persistent complaint that the members of the council is not given a free hand in exercising actual autonomy in their respective blocks. This is due to lack of commitment on the ruling government's part or even the anomalies present in the administrative machinery. Perhaps to put and end to such abnormalities Reangs, the second largest tribe of Tripura, met at a conference in a small village in South Tripura, in December, 1991.[41] Leaders of all political shades resolved to set up a parallel autonomous Regional Council for the Reangs. The movement for autonomy has also begun in the Jampui hills of Tripura. The Mizo convention clamoured for a separate autonomous tribal district council for the seven Lushai clusters of the Jampui hills, and set up a Regional Council for the Mizos of Tripura.[42] These new occurrence show that the demand for separate autonomy status to each tribe is gathering momentum with each passing day. It can be thus concluded that these demands are obviously born out of the failure of the state and ADC administration to accelarate developmental activities.

The aborigenes of the state were getting marginalised gradually, which made them prone to unrest. Granting of autonomous status to these tribes seemed the right answer. The initial years proved quite fruitful as the projects were implemented quite enthusiastically. However, the momentum soon lost the vigour, a growing stratification emerged among the tribals too. The council leadership also belonged to this privileged upper class and the mass rural tribals became victims of structural stagnation, economic immobility and poor entitlement to productive resources. Autonomy therefore, did not quite solve the basic problems of the tribal population of the state. They inhabit the remote hills, in scattered dwellings as a result they do not come together for any realistic movement, rather give vent to sudden isolated revolts. And ultimately these groups make a settlement with the authority on some understanding here and there. One should remember that such moves cannot solve the fundamental economic problems and thus violent unrests continue in the state. That the creation of the Autonomous council has not been able to uproot the problems of the tribals is evident from the fact that the insurgency which had started since the 1950s continues still. The socio-economic disparities, which lie at the core of the insurgency problem still exist in the society. Setting up of Autonomous bodies could not bring under its purview all the tribals of the state or even cater to all their demands. These autochthons who are outside the boundaries do not get the facilities of the autonomous council. Identifying the causes of grievances is a step towards dealing with the people. The state has a majority of plainlander population, so demarcating a large area for only the hillsmen cannot solve the land scarcity. The government' procedure of sympathising with the tribal aspiration and trying to bring them within the mainstream of constitutional political process may appease a certain section, but it creates another schism between the privileged and the non-privileged. Rather, cautions steps should be taken which can combat insurgency like implementing employment schemes, educating these people to avoid wastage of natural resources and instead of creating water-tight departments to deal with the various problems, departments may be overlapped so that all the problems are dealt with by any member of the organisation. However, one point remains the most deciding factor in an insurgency i.e. the population factor. Population of this state as has been seen has increased abnormally, and it is simply not possible to discriminate between the two segments. Yet, infiltrations may be checked further and a policy regarding their settlement needs to be evolved. The autochthons undoubtedly remain at the receiving end, but even the plainlanders are nothing better. So while reserves were opened to accommodate the displaced people, autonomous council was set up for the distraught indigenous people, appeasing one section at the cost of another cannot be the proper answer in dealing with the land scarcity crisis. Autonomy in its actual sense needs to be given with an eye to the economic well being which can ideally bring the economically backward people at par with maintream population. This needs a monitoring cell, which will review the progress of different welfare schemes time and again. Simply allotting land and giving a monetary compensation hardly acts as an incentive to specially those who are used to nomadic way of living.

The concept that the autochthons are becoming a 'minority' should be rooted out completely and a stop be put on the mass scale immigration too. The tribes, on their part need to come together, irrespective of their race, community, sect or political affiliations. Perhaps only then will an egalitarian society be created which will unite the people to fight for a common cause. The state has to come forward and remove the constraints like delays or funding problems in implementation of projects. The sense of belonging should permeate among them and thus the sense of alienation may be wiped out. The benefits, then, can be reaped by all the tribals and non-tribals alike so that no privileged class may emerge representing an aristocratic stratum. For these to materialise a new strategy of development is needed to bring out the remote areas' tribal population in the mainstream. Literacy programmes, awareness schemes, vocational training like weaving, seri-culture, horticulture etc. should be encouraged which would remove ignorance from among the people. These steps would undoubtedly remove the barriers in development process and improve their status thus enabling the indigenous people to exercise their free will.

NOTES AND REFERENCES

1. *The New Encyclopaedia Britannica*, Univ. of Chicago, U.S.A., 1989, Vol. II, p. 775.
2. Govt. of India Act, 1935: Sec. 311 (1)–Areas along the Frontiers of India and Baluchistan which are not a part of India or Burma, or any Indian or foreign state.
3. Hunter, W.W.: *A Statistical Account of Bengal,* Vol. VI (1973), p. 510.
4. Mackenzie, A.: *History of the Relations of the Government with the Hill Tribes of the North-Eastern.* Frontier of Bengal, Calcutta 1884, p. 275.
5. Sinha, K.C.: *Rajmala,* 1856, p. 122.
6. Sharma, R.M.: *Political history of Tripura,* 1986 Calcutta, p. 157.
7. Singha, K.C.: *Op.cit.,* p. 355.
8. Mackenzie: *Op.cit,* p. 67.
9. Roy Burman, B.K.: *Demographic and Socio-Economic Profiles of the Hills of North-East, India, Census of India,* (1961), p. 372.
10. Chunta Prakash (extraordinary issue), Jan., 16, 1930.

11. The Tripura Tenancy Act. (in Tripura Code Part II, containing the old Acts of the rules of the state.
12. Tripura State Gazette Sankalan, TSA pp. 135-37

    Advisory Council–31st August, 1927.

    Legislative Council–3rd Sept., 1927.

    Executive Council–17th May, 1929.
13. *Ibid.*, p. 8.
14. This sort of development society was described by La Palambara as–"It is surely literacy that will be virtually instrumental in bringing peasants to understand that they can achieve greater control over their physical environment: La Palambara Joseph: "Distribution: A crisis of Resource Development" in Binder, Leonard et al., Crisis in Sequences in Political Development, Princeton University Press, 1971, p. 269.
15. Sen, T.C.: *Tripura in Transition,* 1923-57, Agartala, 1970, p. 3.
16. Bhattacherjee, H.: *Communism, Nationalism and Tribal Wuestion in Tripura,* EPW Vol. 35 (Bombay) No. 39, Sep., 29, 1990, pp. 22.12.
17. T.S.G. extraordinary issue, 15.10.49, p. 4.
18. Ganguly, J.B.: *The Economic Content of the State Formation Process in Medieval Tripura,* 1968, p. 151.
19. Case of Land at Champamura–No. F. 4 (2A-Rev/59) Tripura Secretariate Archives. (Unpublished).
20. Touji No. 341-No. F. 412 A.–TSA (unpublished).
21. Dev, Dasarath–Tripura Ganatantrik Andolone Ganamukti Parishader Bhumika (Akansha) translated from Bengali article in Souvenior of Tripura Employees co-ordinations committee–5th Biennial Conference Feb., 13-16, 1981, p. 4.
22. Dev, D.: *Tripura Upajatider Bhumi Sanrakshane Upajati Ganamukti Parishad* (Translated from Bengali article) *op.cit.,* p. 2.
23. Memorandum of the Tripura Rajya Gana Mukti Parishad: Sep., 10, 1955 submitted by D. Dev. to Prime Minister J.L. Nehru.
24. An article by J. Debbarma in the Daily Desher Katha (a news paper)–Agt. Mar. 24, 1983 (Translated from Bengali).
25. *Administrative Reforms Committee: Report of the Study Team on the Union Territories and NEFA*, Delhi 1968, para 691.
26. *Tripura Legislative Assembly Proceedings* 22.3.1966, pp. 30-34. A Resolution by Sudhanwa Deb Barma.
27. *Tripura Legislative Assembly Proceedings* 15-3-67. Ammendment moved by Aghore Deb Barma pp. 27-28.

28. *Tripura, Legislative Assembly Proceedings* 15.12.67 (Private members' resolution by A. Deb Barma and supporting speech by Abhiram Deb Barma and B.C. Deb Barma), pp. 28-33.
29. *National Committee Report on Land of the Tribals,* 1981.
30. Forest Department Notification, No. 13.29.4.1952, Govt. of Tripura.
31. *The Tripura Tribal Areas Autonomous Council Act,* 1979, Tripura Act. No. 12 of 1979, Tripura Govt. pp. 3-13.
32. *Annual Report of the TTAADC*–Three years of work 1982-85 Agt. p. 12.
33. Report published by the TTAADC of their work in the last ten years, p. 36.
34. *Ibid.,* p. 48.
35. *The Tripura Tribal Areas Autonomous District Council Act. op.cit.*
36. *Ibid.*
37. *Ibid.*
38. Bhattacherjee, H.: TUJS alienated from tribals, *The Statesman, Wed,* July '93.
39. Chakraborty, S.: *Politicians make a sport of tribal issues.* The Statesman, 26 Jan., 1995.
40. An article in Syndan Patrika, Tripura 12th Dec., 1991.
41. *Ibid.:* 20 Jan., 1992.

8

Land Alienation

Land alienation as is understood is a dynamic concept and does not mean a mere transfer of land from one to another. It is the balance between loss of economic right of a lower order and gain of social power and economic efficiency of a higher order. This concept of land alienation was unheard of even in the early 20th century. It is a recent phenomenon but its harmful consequence can be already felt in a primarily agrarian economy. With the growing dependence upon technology, and accelerating development in the economy there is an increasing diversification of the economic pursuits too. Land, therefore, becomes a marketable natural resource its demand increases and its availability raises concern. Under such strains, the hitherto actual users of land quite often have to relinquish their hold on land or even give up the hope of acquiring land further and thus 'Land Alienation' sets in. This is the subject of the present chapter and is seen as the core of the study. A brief background of the land systems of North-east is discussed to get a picture of the raging problems.

The North-eastern sector of the country, as has been already discussed, faces various impediments in regard to land availability the first and foremost being natural. The hilly topography obstructs the spread of settled cultivation and added to this there is the extraordinary growth of population in some states specially Assam and Tripura. Moreover, this region, one should remember is primarily inhabited by various Indo-Mongoloid tribes, who believe in communal ownership of land. Individuals who claimed to acquire right of use of land could do so only at the Chief's recognition. Absence of individual ownership pattern forbids the knowledge of actual proportion of land alienation. The land tenure system of the North-east, is therefore, inadequate to give a perfect understanding of this rampant problem. Modernisation and development of agriculture requires spread of settled cultivation which necessitates the individual possession pattern, and a close study of the same shows that quite a considerable area of arable land has already passed into the hands of rich non-agriculturists. A new land-holding class has emerged, which though not the actual users, live out of the same dispossessing the hitherto users of land.

Arunachal Pradesh is inhabited by more than hundred tribes, who mostly practise community ownership of land. The Balipara Frontier Tract Jhum Regulation, 1947 (Regulation III, 1947) provides for land under permanent, semi-permanent cultivation, dwelling houses are subjected to clan or individual ownership, otherwise vested with the state.[1] The individual holdings can be transferred, sold or even alienated but only within the village premises, not to any outsiders. In the hill districts of Assam, private ownership of land is unheard of while various leases are given away on the plains.[2] The latter creates veritable rights, interest, which were simplified by converting annual leases to periodic leases. The Assam State Acquisition Act[3] (1951) was promulgated to acquire the rights of proprietors and intermediary tenure holders, and give the tenants the status of land-holders directly under the government. Manipur is one of the states where ryotwari system of land tenure is in force, though only on the plains. By the enactment of the Manipur Land Revenue and Land Reforms Act, 1960, the rights of each holder is recognised by law which acts as one of the safeguards against land alienation. The land tenure system in Meghalaya is quite complicated, the individual ryot does not have absolute ownership rights over the land cultivated by him, though land under settled cultivation is virtually owned by the family concerned.[4] Sale, transfer or mortgage if done at all, has to be within the clans. In Mizoram, shifting cultivation is associated with communal ownership of land, individuals do not have absolute ownership right the chief enjoys autocratic powers in distributing land among the villagers.[5] However, recently land is being allotted to individuals for terrace cultivation, orchards[6] etc., but these are still quite insignificant. In Nagaland, ownership rights are determined by tradition and customary laws, though individual rights on terrace land is recognised which is again permanent, heritable and transferable though within the tribes.

Backdrop to the Land System of Tripura

The princely state of Tripura, as seen in the foregoing introductory chapter comprised of the fertile parts of present Bangladesh, vast tracts of hilly land extending upto the Lushai country east of Manipur, as well as the narrow hilly tracts running between Hill Tipperah and Manipur.[7] Agriculture constituted the chief occupation, and land availability did not pose any constraint initially. However, area of the state gradually dwindled with the slicing off of various peripheral land, British occupation of state territory, exigencies along the Lushai borders but even after troubles subsided they failed to return the occupied region. The demarcation of boundary at Langai river also claimed quite a considerable area, as the British contemplated that a river may serve as a definite boundary, since hills could be occupied by shifting cultivators but river banks would be left unoccupied as plough cultivation was not yet quite popular.[8] The areas taken away by the British were undoubtedly fertile tracts which could have been put to plough cultivation were they restored.[9] Thus land availability began to shrink, and the pressure of British revenue exactions rose, the kings went on inviting the neighbouring plainlanders, giving away precious arable land, homestead land, at nominal rents. The revenue returns excelled over those of the shifting cultivators and even by the end of the 19th century or in the beginning of the present century this process did not create any misunderstandings. The major political event which created panic amongst the Hindu population of erstwhile East Pakistan was the partition which brought along with it an influx that changed the entire history of the state. Scores of plainlanders poured into this land in search of peace and security as Tripura already bore an ethnic affinity towards them. These immigrants were mostly of the cultivating stock and 'settled' on whatever land they could lay their hands upon. Now began a steady, fast decline of available arable land, that posed a threat to the land question of the state. As the saturation point of absorbing the immigrants was reached, pillars were erected along the borders on the basis of authentic documentary evidence and land records[10] but the influx continued even in the 1970s and 1980s (Bangladesh liberation war, Chakma atrocities) and is continuing unabated still, though under covers. As already seen the indigeneous masses preferred the shifting mode of cultivation though settled cultivation was not exactly unknown to them. The immigrants, viz., the Bengalis, the Manipuris and the Cacharis took over ploughing, but land posed certain natural impediments (80 per cent of land is hilly and 55 per cent of cultivable land lies in the hills). In the face of such bottle-necks any attempt to develop agriculture seemed rather tough, and the land system also proved inadequate.

A look into the chart (p. 137) will enable one to understand the prevailing land system of the princely regime[11] and serve as background to the land question of the state. There were no hard and fast land tenure rule in the past, the king was the superior landlord (Zamindar) though he actually retained very little under his management. Most of the plains land (Taluk) were granted by the king to his intermediaries (Talukdars) in perpetuity at fixed rents. There is no definite information about land ownership among the original inhabitants. The King was the owner of the hills too and leased out the land to tribal chieftains (Thakurs) who in turn distributed the same among his tribesman. An assumption served as customary law here, land belonged to him who first tilled it. They preferred shifting mode of cultivation and owned land communally, no outsider could lay claim over any plot of land without the permission of the chief, so in the 19th or early 20th century there was no question of land transfer. But with the coming in en masse of the neighbouring plainsmen, this land was opened up gradually, the shifting cycles got reduced, the available land shrunk resulting in loss of fertile soil which led to meagre crop production, hence the self-sufficiency aspect of the tribals seemed to be on the verge of break-up. Those who had the means moved on to settle on the plains and others retreated to remoter hills. The ruler

Chart : The land system on the plains

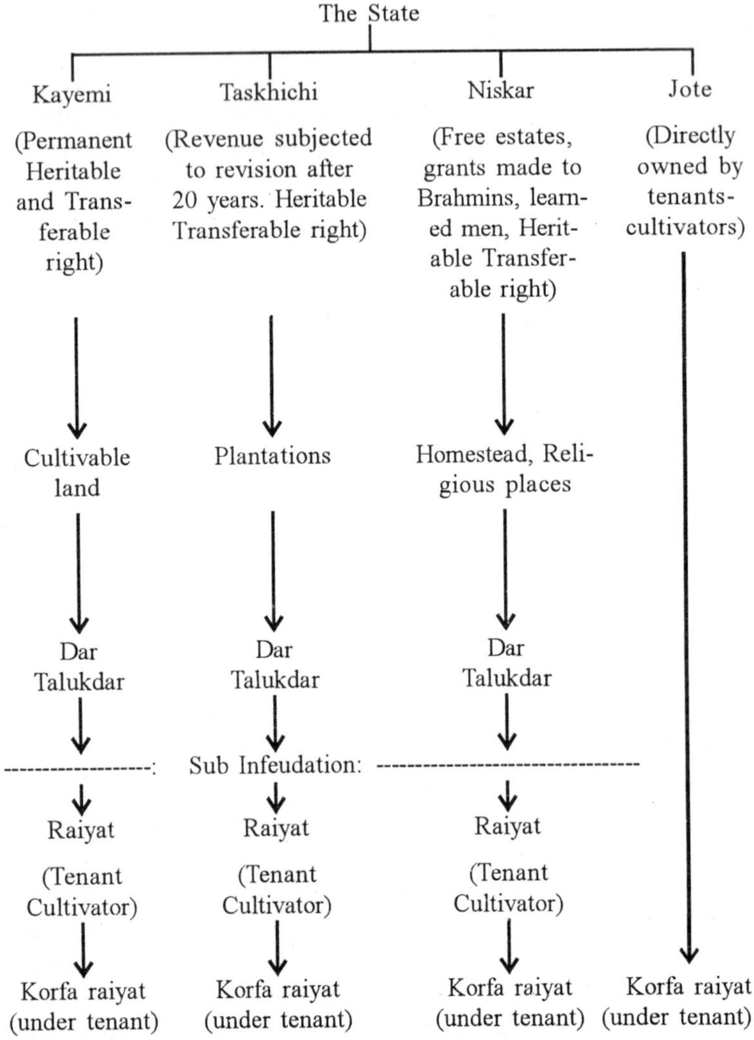

owned the entire land of the state but he made grants of the land on the plains to three intermediaries (i.e., landlords or Talukdars). They were Kayemi talukdars who were given land for cultivation purposes, Taskhichi talukdars who owned plantation land and the Niskar talukdars who were granted revenue free land for religious purposes or homestead. Sometimes land settlements were also made directly with raiyats (actual farmers). These landlords leased out their lands to the actual farmers or sub-landlords known as Dar-talukdars, who were intermediaries between the talukdars (landlords) and the actual tillers. But these Dar-talukdars again sub-leased their lands to raiyats whose right on land was heritable—not transferable without his landlord's permission, but his ownership rights could be sold off if he failed to pay rent by due date. To avoid such dismal circumstances the raiyats often sub-let a part or the whole of his holdings to korfa tenants or under raiyats. These under-raiyats (sub-tenants) enjoyed jote rights of the raiyats' land on payment of rent generally in the form of produce Adhi, bhag, barga etc. As is evident, the tenurial pattern in Tripura during the past did not provide ownership rights. Moreover, presence of several intermediaries due to infeudation and sub-infeudation posed immense land related problem. Land system structure resulted in fragmentation of rentier interests and ultimately led to the relatively new concept of land alienation. The king always tried to attract plough cultivators for profits as is evident from the following excerpt of Shashan Sammelani of 1913 A.D.[12] "....the system of permanent settlement should be continued. Instead of distributing small pieces of land, large area of 500 drons (1 dron=16 kanis) or more in one block should be given out in permanent settlement. This measure will lead to an increase in the number of rich and cultured persons as also ordinary tenants". This view of the kings of the state echoed the basic concept of the law of Landlord and tenants, 1886 (Praja Bhumy-Adhikari Sambandh Bishayak Act, 1296 T.E.)[13] which stated that each tenant pay fair and equitable rate of rent determined by the prevalent rate and the value yield from each kani. The tenant farmers received 25 per cent of the profit and the landlords were ensured 75 per cent. The Tripura Tenancy Act of 1886 A.D.[14] (amended in 1945 A.D.) promulgated that a tenant may acquire occupancy right by continuous possession of the plot of land for seven years, provided rents are paid regularly. It also said that a person may enjoy the income by settling his lands with a tenure holder but such a tenant does not have the status of a raiyat. But as the pressure of population increased anomalies in the contracts were apparent, the existing tenants were often evicted even before seven years term was over on some pretext or the other and another tenant recruited in his place who was willing to cultivate the land at an inferior right but a higher rent, following stricter rules. These acts, however, kept silent over the Korfa tenants rights (the under-raiyats). There are no documents to show that the tenancy is one at will and that the tenant can never acquire occupancy right or ever be evicted. Even the decisions of the High Court of Tripura remained silent on this fact.[15] Official records found at the Archives states that, to protect these tenant cultivators, whenever the state tenancy laws fell silent, the Bengal Tenancy laws, 1885 were applicable. (There was a provision that the under-raiyats or bargadars could not be evicted except under the C.D.E. clause of the Bengal Tenancy Act, 1885).[16] But how many of them actually knew of these provisions? The practice of the landlords selling their lands occupied by their tenants were rampant.[17] If these laws were implemented ably could there be so many landless labourers in such a short span? A continuous process of mortgaging, sub-letting, selling and transferring of the raiyati interests resulted due to the inadequacy of the land system in the princely regime. This process of continuous settling, re-settling of the agricultural population led to mass scale immigrations by the turn of century. The rulers aimed at better utilisation of the land so that revenue may increase as well as an all round development be achieved. It was found that a substantial number of tenant cultivators were attracted, not big landlords who remained mostly absentee holders of taluks and realized rent from the tenants by letting out their 'taluks' in small plots.[18] If this is taken to be the order of the time, it may be deduced that whole of the state was mostly inhabited by small landholders or tenant cultivators (the causes of their immigration has been already dealt with in chapter 3). Since the pressure on land still remained low small taluks were often given away at nominal rent and the cultivator virtually enjoyed the position of a peasant proprietor, where in the average size of farms was around 5 acres at least till the early 20th century.[19]

Initially, the indigenous population stayed away from joining the race for land grabbing because they still shied from plough cultivation and though the problem of land alienation could not be perceived yet on the plains these people sufffered much as the newcomers settled on the foot hills and low hillocks even. They lost hold on their land which they never did 'own' and turned labourers on the field still not being aware of land as a productive resource, and even the state did not bother to treat them at par with the settlers, the kings were content with the revenue returns and the increase in the total cropped area. Thus acres and acres of land passed into the usufructuary occupation of smaller units (family or individual) which was once in a common possession (clan or village). The continuous migrations from the erst-while East Pakistan, giving away of large tracts for their rehabilitation and there being virtually no improvements in the techniques of cultivation, land was the only means of sustaining the growing population. A new landed class emerged, as land got divided and sub-divided continuously. It was at this juncture that the most serious land related problem came into light–Land alienation.

Causes of Land Alienation

Alienation of land among the original inhabitants of the state was greater. Belief in age-old customary law gave them the conviction that land belonged to him who first cultivated it and so did not feel the urge to record their ownership rights. Rather, they carried on shifting to establish their right to 'possess' the land which had no legal base. So whenever the question of legal owner of a plot arose, the claims of the autochthon was side-tracked. The evil effect of shifting was felt when the top soil got eroded. Apprehending ecological disbalance and further loss of valuable revenue, shifting cultivation was restricted on certain belts by government orders. Those involved turned to the forests, but here too they had to face discrimination, 'Jangal Abadi' Schemes were attracting the settled cultivators in great numbers, as the kings offered the benefit of concessional rates of rent till the twentieth year. The tribals' entry into the forests were also restricted, they now had to seek livelihood as labourers giving up any hope for acquiring land. Coverted moves of the non-tribals also dispossessed many autochthons of their land. The gullible autochthons were often coaxed into mortgaging their land to the settlers on the plea of putting it to proper utilization. But it was seen that even after the lease period was over the land was not restored, rather the mortgagee usurped the land dissuading the mortgager from filling any applications for the same. The ignorant hillsmen gave in as they did not have the necessary means to pay for revenue arrears or the mortgage fee even and the land changed hands under covers. This phenomena acquired great proportions during the period following partition, when the state government took up rehabilitation of the displaced persons to be their moral duty. Land which legally belonged to the tribal masses was given away without putting forth any effective measures to protect them from such dispossession. Though about 1760 sq. miles area was reserved by the kings for five indigenous tribes (42.07 per cent of the population) there were many who lived outside these reserves and their claim to land went unheeded, these people were more prone to be dispossessed of their land as they lived in close proximity to the settlers. An example of such land alienation may be given here– A khas land in Khowai belonged to a tribal peasant who had reclaimed it after much effort. This land got marked as to be fallow land with bushes growing on it, and so was allotted to a displaced person with no compensation being made to the actual owner.[20] However, one should bear in mind, that the refugees who came to settle in the state were not much better off either. They were not quite familiar to the laws of the state and so could be easily duped by the big land holders. They were driven out of their homelands and held onto the last straw for survival, so when they received any plot as tenants even they did not bother to look for legal documents. At the slightest fault, the landowner evicted him and brought another tenant who was even more eager to repeat the previous owner's 'mistake'. These settlers at times found themselves evicted even after receiving government land. An example may be cited here, a refugee purchased 2 kanis of land at Jogendranagar (Jote No. 234) from the Rehabilitation Department on mortgage. But soon it was found that the government had rehabilitated some other refugee families on the same land.[21] These were no doubt discrepancies on the part of the Refugee Rehabilitation Department, but they were only the beginning of a long drawn problem. Undoubtedly the displaced people utilised land to its optimum capacity and so the department was eager to settle them. Agriculturist families were settled in a portion of land at Hatileta (totalling 12.80 acres) a dense jungle area which the refugees cleared and reclaimed. 2.00 acres were given for rehabilitation at first and then 1.90 acre and 0.90 acres of land was released for the same. These were fallow land lying in the jungles but were used by the displaced people for homestead jute cultivation.[22]

Moreover, tea cultivation was found to be quite congenial to the climate of Tripura. This could make lucrative business for the ruling elite, as they enviously looked at Assam's high profits in this trade. Almost 5555 sq. miles of slope land were leased out. This could also employ a number of landless peasants as coolie labourers, but the hillsmen could not be lured to this laborious, unfamiliar job and soon deserted the plantations. This too was a kind of alienation, for land which could be subjected to shifting was utilised for another plantation not catering to the needs of the indigeneous people. A mention should be made here of Chakla Roshanabad, a Zamindari held by the Maharaja, the impartible part of the state of Tripura and from which a considerable amount of revenue could be derived (in fact more than the state). After the merger of the state of Tripura this 555 sq. miles constituting the most fertile area was awarded to Pakistan. If this land could be treated as the king's own private property perhaps land scarcity would not have been so glaring.[23]

The various causes of land alienation in the pre-Independence era gradually gave rise to a privileged section both amongst the original inhabitants and the settlers. The unsuspecting simple cultivators were evicted from their own land, due to the anomalies in the land systems or the Rehabilitation Department's policies and even if they were compensated hardly anything percolated down the ladder of middle-men, which could suffice for the massive loss of land. This process increased manifold after Independence and Tripura's accession to the Indian Union in 1949. There was an unprecedented population increase, vast areas of vacant government (khas) land were unauthorisedly occupied which considerably altered the configuration of the fields as well as ownership of the settled land due to inheritence, purchase, partition etc. The land system obviously failed to meet the requirement of the new situation[24] and an overhauling was needed. With the pressure of population on land rising, the self-subsistent economy of the hillsmen was giving way to one of a wider kind. As this economic transformation was setting in the hillsmen faced a dilemma, whether to seek refuge in remoter hills or change over to plough cultivation which was undoubtedly quite profitable. The elite among them (the chiefs/the Thakurs) did change over to wet rice cultivation. They now came in contact with the new settlers and learnt about the amenities of social life. They also learnt the facilities of money economy, got consious of what money could buy and embarked upon the habit of borrowing even. Hence started another new cause of land alienation–Indebtedness. The Report of the Political Agent[25] (1878-79) reveals that the peasants of the plains were free from the curse of indebtedness, but the hill people, almost as a rule were always indebted to the Bengali money landers. They were always in a pathetic condition, improvident, idle and addicted to drinking, knowing no skilled labour and worked as day labourers but contracted debts from which they could not free themselves easily. It was reported that with the coming of the neighbouring plainlanders the entire trade and commerce of the state were controlled by them who gave loans readily in lieu of huge interests. Getting exposed to the new standards of living the noveau-riche among the hill people spent lavishly on clothes, jewellery, festivities and thus ended up in the clutches of the unscrupulous money-lenders. Obviously owing to hand-to-mouth conditions the borrowers could not repay, their produces merely sufficed for their own consumption, being entirely new to this mode of cultivation and therefore had to pay with their land. Such consumption loans accumulated over time, involving a virtual hypothecation of crops and ultimately land even. It was mainly the marginal and the small farmers who suffered irrespective of their race or ethnicity. The terms were onerous, unfavourable to the agriculturists wherein the conditions of borrowing are such that a big share of the income is taken up by interests only and debts are often repaid by further debts leading to ultimately alienation of the land.[26] Moreover the Dadan system (receving money in advance) initiated during this time rendered many hill tribes landless. Quite a large number of these hill people borrowed money, rather took an advance from Bengali traders at exorbitant rates of interest on condition that within a certain time, bamboos, timber or other forest produces (of specified quantity and quality) would be given in return. But when they failed to return it, they had to sell off their produces to the traders at a much lesser price (as fixed by the latter). It is known from the reports of the Political Agent that cotton was the principal source of income but this too they could not sell to their advantage for being under obligation incurred due to the dadan system, but still they are unambitions and not anxious for a better condition.[27]

The peasant money-lender nexus could also be noticed elsewhere, where in both the communities fell easy prey. This was in the case of marketing of products, whenever the peasants wanted to sell his produce in the market he had to seek the help of the village trader (usually the money-lender), being ignorant of the market forces, price complexities, transportation etc. Mostly, this trader offered his services as the buyer of the produce and paid a price which was much lower than the prevalent market rate. Yet, they were most trusted by the peasants because at times of distress, they helped these poor masses who were unknowingly tied in bondage. So any chances of repaying the borrowed amount through payment was blocked. The tendency of the peasants borrowing at high rates and selling their produces to the local trader (money-lender) at lower rates simply entrapped them in a vicious circle which ultimately led him to part with a portion of his land. This confirms the fact that the dominant trend of agrarian transformation is towards the emergence of a new antagonism between a well-to-do landlord or money lender trader class and the increasingly impoverished mass of small peasants, labourers or share croppers.[28]

Land utilisation schemes underwent a few changes in the post accession era. Apart from providing land for cultivation, homestead, rehabilitation, the government had to provide for developmental purposes too. This transition from an essentially rural economy to an urban one meant building roads, hydroelectrical projects, houses, schools, offices, hospitals, military outposts, civic amenities etc. A network of roads were built by acquiring land from the people, the Assam-Agartala Road the Agartala-Sabroom road even roads built on the Indo-Bangladesh borders. Tripura, being a border state, security acquired a very important place for accommodations the CRPF camps, BSF and TSR camps. In 1952, lands were acquired under Act I of 1894, Land Acquisition Act for the extension of the Tripura Military Cantonment.[29] However, records show that the owners were kept in the dark regarding this acquisition, and when the government passed an order to release the same and restore it to its owner, the Military department looked the other way.[30] There is evidence that private land has been acquired on permanent basis for various purposes by paying a fixed compensation.

These show that with the passage of time as more and more land was being utilised for number of reasons the resource available for cultivation purpose was getting curtailed. Land alienation acquired huge dimensions as the marginal farmers, the independent jhumiyas were deprived of their sole means of survival, and following partition there is a constant increase in the landless agricultural labourers number as is evident from the Table 8.1.

Table 8.1 shows that during the decade following independence (1951-61), scores of refugees settled here, they served the state with increased crop production as they spread extensive settled cultivation hence the huge number of cultivators. But in the following decade (1961-71) the picture changed altogether, the number of cultivators shows a marked decrease as the agricultural labourers increased. The trend continues even in the following decades as influx gather momentum, land available shrinks due to several other reasons (already discussed). However, during the decade 1981-91 it can be seen that both the number of cultivators as well as landless labourers have fallen. The reason for a decrease in cultivators can be well understood but that in landless labourers accounts for the fact that more and more such labourers are turning to urban jobs or plantation labour. This gives us a clear understanding as to how land gets alienated from the actual owner turning him landless. The peculiarity in Tripura's case is that the hillsmen have to bear the brunt of such land alienation being averse to plough and giving more importance to age old shifting mode of cultivation. The new settlers received government's attention more. Consequently, a section of the old elite, who were formerly in the service of the kings of Tripura found themselves losing grip over the politics of the state. In their place, the plainlanders emerged as a socio-political force who wielded power too and associated themselves with the political leadership of the state.[31] The state came to be divided on communal lines as the former instigated the poor peasantry over land ownership rights.

Table 8.1: Trend of Cultivators and Agricultural Labourers in Tripura (As a percentage of the total population)

Year	*Cultivators percentage*	*Agricultural labourers percentage*
1951	62.94	8.93
1961	64.24	77.53
1971	54.41	19.26
1981	43.57	23.91
1991	29.09	11.08

Source : Census of India, Various issues.

Steps Taken to Alleviate Land Alienation Problems

The state government did not turn a blind eye to this new transition. Much concern about the plight of the peasants was shown who had to forsake their hold on land. The contemporary administrative records show that compensations were duly made on land acquisition but doubts are raised whether it reached the peasant at all. To ease their burden, the Tripura Land Revenue and Land Reform Act was passed in 1960.[32] It aimed at abolishing all intermediary interests on land and bringing the actual tiller of land in direct contact with the government. All the old laws and tenurial patterns were abolished, even the tribal reserves of the princely regime were done away with. The act helped in wiping out the socio-economic discriminations to enable both the segment of the population to enjoy optimum returns on his labour and capital investment. It also aimed at imposition of land ceiling, preparation and maintenance of record of rights and tried to put an end to land alienation. The foremost and top priority of all was given to the land belonging to the hillsmen, no land of theirs could be transferred to a non-tribal without the permission of a Magistrate/Collector/ Sub-Divisional Officer. Now all the peasant cultivators became a raiyat directly under the government with permanent rights on land he even had the right to create sub-tenants (Korfa raiyat).[33] The original holders of land were paid compensation as was admissible according to the provision of the Act. Initially the compensation was paid by the government on a declaration that it should be realised from the raiyats by annual instalments. He could lease out land to another person on such rent as may be agreed upon between him and the person subject to the maximum rent provided in the Act.[34] This lease could be heritable but not transferable. Another provision in the Act was imposition of land ceiling at 25 std. acres for a family of five and an additional 2.08 ha. for each excess member to a maximum of 20.23 std. ha. homestead land, plantation, colonies were excluded from the ceiling. Partition of holdings was restricted into less then 2 std. acres. Any person acquiring a right on land was required to report of his acquisition to the village tehsildar within three months' time.

These provisions of the TLR & LR Act, (1960) tried to aim at controlling the raging land alienation problem. It made clear a fact that land shall not be deemed to be under the personal cultivation of a person unless he engages in the principal agricultural operations of the said land. The results initially were quite encouraging, the estates of 29.215 intermediaries in the state were vested with the government and the former compensated duly. Thus the Act helped in removing the evils of feudal land relationship also. But the inequalities in the distribution of land holdings could not be reduced, nor could it provide security of tenure to the under-raiyats or stop rack-renting which took place covertly. The zamindari system was abolished undoubtedly, there were no particularly big landlords, but the land ceiling had quite a few loopholes. Firstly, in the absence of land records, it was found that the jhumiyas "possessed" quite a good deal of surplus land. On the pretext of imposing land ceiling these land were distributed among the landless (both the tribals and the settlers); secondly, the minimum limit (25 std. acres) was doubled up by the landlords under disguised identities; thirdly, the Act promulgated that 1 std. acre was equivalent to 1 acre of lunga or nal land or 3 acres of tilla land–the land-holders showed plain land as tilla land so that it could be kept under a single family's personally cultivated land; fourthly, land for non-agricultural purpose was exempted from ceiling limit which gave immense scope for circumventing the provision. Thus actual land supposed to be distributed among the landless was found to be much less as is portrayed in Table 8.2.

Table 8.2: Size of agricultural holdings and their area

Size of the holdings (ha)	*Numbers*	*Percentage of total holdings*	*Area (in ha.)*	*Percentage of total area*
0.00-0.5	1,14,846	46	27,114.30	11
0.5-1.0	58,944	24	43,056.41	17
1.0-2.0	47,049	19	66,167.08	26
2.0-3.0	15,976		38,506.44	
3.0-4.0	6,242		21,359.19	
4.0-5.0	3,043		13,536.83	
5.0-10.0	3,409	11	22,130.29	46
10.0-20.0	509		6,382.31	
20.0-30.0	37		881.59	
30.0-40.0	10		335.30	
40.0-50.0	15		625.91	
50.0-& above	54		13,981.01	
	2,50,134	100	2,54,076.66	100

Source: State Report of World Agricultural Census 1970-71 in Tripura

Table 8.2 shows a huge increase in the number of marginal holdings as a result of the implementation of the Tripura Land Revenue and Land Reform Act, (1960). A plausible deduction may be made here, those who were "settled" on government khas land since the 1950s did not really cultivate their land but leased them out in small holdings to share croppers, sub-tenants, who being landless readily agreed to till the land even without any legal documents or contracts. Insecure tenures could easily victimise the tenants and sub-tenants. The feudal legacy of the rural power structure also seemed to be in the favour of landowners and money-lenders. Hence the inequality in land distribution, while 46 per cent of marginal farmers operated 11 per cent of the total operational holdings, 11 per cent big landowners were comfortably operating 46 per cent of the holdings. Land transfers could not be stopped, but there was lesser gusto at the moment which now took place beneath covers due to the restrictions of the TLR and LR Act, 1964. Obviously alienation of land could not be stopped the big landowners became absentee landlords enjoying greater advantage over various related facilities (e.g. irrigation, fertilizer, seeds, loans etc.) while the marginal farmers remained vulnerable as before if not more.

Soon the need was felt for a Second Amendment (1974) which aimed at providing for restorations of "illegally" transferred land. To prevent unauthorised occupation the cut-off date was fixed on 1st January, 1969 i.e. henceforth if any land belonging to a tribal was occupied unauthorisedly by a non-tribal the latter was liable to eviction by the Circle Revenue Officer.[35] But, those transfers which took place before 1st January, 1969 did not come under the Act's purview. The ceiling limit was scaled down to 7.20 std. hectares. The amendment envisaged a square deal to the underprivileged agrarian class as a whole. To stop rack-renting, it also laid down the fact that the under raiyats had to surrender their land to the collector and not to the raiyat. This did have a certain impact upon the land availability factor as quite a vast area was released after the ceiling imposition, as depicted in Table 8.3. The Table shows a noticeable increase since 1970-71 in the number of marginal holdings which rose from 1,14,846 (1970-71) to 1,44,897 (1976-77) while the large holdings measuring to hectares and above fell from 625 (1970-71) to 605 (1976-77). But quite a noticeable fact here is that there is a comparative increase in the size of large holdings which highlights land alienation again. It may also portray another aspect of the process, that land was being "owned" by the same family furnishing fictitious names and identities. Enquiries reveal that even under-raiyats, tenants loyal to the landowners were taken into confidence and land purchased by the landowners in the former's names (the entire land actually belonged to him who paid for it). Initially the number of operational holding did increase as a result of such measures, from 2,50,134 (1970-71) to 3,07,700 (1980-81).

Table 8.3: Agricultural holdings after ceiling imposition

Size of holding	*Number*	*Percentage of holdings*	*Area (hs.)*	*Percentage of of the area*	*Average size*
0.00-1.00	1,44,897	58.8	66,441	21.5	0.46
1.00-2.00	60,112	24.4	85,781	27.7	1.43
2.00-4.00	32,893	13.3	88,116	28.5	2.68
4.0-10.0	810	3.3	43,515	14.1	5.37
10.0 and above	605	0.2	5,293	8.2	41.81

Source : 1976-77 Agricultural Census, Statistics Department, Agartala.

A change of 24.80 per cent was noticed, as well as a decline in the number of average holdings.[36] So to escape government's pressure, in reality loopholes were sorted out and new rules fabricated but alienation could not be stopped. One cannot expect that such pressures would lead to any spate of retransfers of land which has already been alienated. The number of landless peasants have also increased, yet the new ceiling limit did deliver some surplus land (3247.75 acres). By 31st Oct., 1977 the government took possession of 830.34 acres and distributed 192.49 acres among the landless and marginal peasants (46.30 acres to 47 scheduled caste peasants, 3.48 acres to 3 tribals peasant and 142.71 acres to 157 other peasants). In the 1980s, 2,011,65 acres of surplus land was received of which 1916.146 acres were taken by the government and the remaining 95.41 acres had to be forgone being involved in legal discrepancies. Of these 1916.246 acres 209.36 acres were given to 235 scheduled caste peasants, 425.91 acres were given to 313 scheduled tribe peasants and 883.32 acres given to 754 other peasants (totalling a distribution of 1578.59 acres). Till April, 1998, 780.66 acres of land was declared land ceiling surplus of which 369.01 acres were disbursed among the peasants of the state. Of this 160 scheduled tribe peasants received 161.80 acres, 125 scheduled caste peasants received 65.17 acres while 142.04 acres was given to 178 peasants belonging to the 'other' category. Analysing these statistics, it is revealed that though the government acquired 3484.956 acres during the 1970s and 1980s it merely disbursed 2080.09 acres.[37] The remaining acquired land was taken up, perhaps, by developmental projects essential in an urban economy. Yet one point comes clear while only 3 tribal peasants availed of 3.48 acres in 1976-77 about 313 such peasant came forward for 425.91 acres of land in the 1980s which portrays the fact that the autochthons are undergoing a process of transition (as described in chapter 4). But again the number fell to 160 only in the latest figures showing plausibly the growing indebtedness of these marginal farmers. The land was obviously not enough for the landless peasants of the state who were victims of land alienation through coercion, indebtedness, or faulty governmental machinery. Moreover, political pressure forced land to be given to unauthorised persons irrespective of whether they were victims of land alienation or not. At the same time another 230 acres was taken away for military bases. That there were discrepancies in land allotment can be seen in a report of land alienation cases revealed by the concerned department, as on 31.8.1986. There were as many as 19,089 petitions deposited in the office involving 18,840 acres of land, wherein land was restored physically to only 3,633 petitioners involving 3,226.51 acres.[38] Keeping in mind that land alienation is much more widespread than what is revealed officially, the above numbers are enormous. Surplus land was distributed, undoubtedly, but a look at the distribution process depicts that the 'other' category received much more than the poor peasants. The latest figures of land allotment[39] in West Tripura shows a remarkable achievement 14,027.35 acres of land is distributed to 6113 landless and homeless tribal families; 8729.52 acres distributed to 4476 landless tribal families and 233.11 acres of land given to 1376 homeless tribal families totalling 22989.98 acres of land given to 11,956 such families. Again, 1297 scheduled caste landless and homeless families received 1326.62 acres of land, 1656 landless families received 154.50 acres of land and 1685 homeless families received 213.67 acres of land, a total of 3694.79 acres given to 4546 scheduled caste peasant families. Moreover, some 3067 landless and homeless families belonging to the 'other' category received 3417.30 acres of the land, while 3293 landless families of the same category received 2416.48 acres of land and 3686 homeless families received 504.40 acres of land, a total of 6338.18 acres of land given to 10,046 families of the 'other' category. Or in other words a total of 43,022.95 acres of land have been distributed amongst 26,548 landless and homeless families in the west district since the inception of such allotment programme. One point comes out clear that though in the initial stages the 'settled' population were pampered while the autochthons get preference nowadays in such allotment of surplus land by the concerned departments. Such restorations hardly solve the problem of land alienation. Alienation of land is a dynamic process and this land often ends up once again in the possession of traders, money-lenders or big landowners through the process mentioned earlier (cause of land alienation). It is also noticed that these peasants often refuse to take back the restored land and cultivate it because they dislike to break the bond of trust and confidence between them and the transferees. They consider this act to be morally and ethically wrong, clinging on to their age-old traditions and beliefs. Or, perhaps they are afraid of incurring the wrath of the big landowners, fearing reprisals as the latter (transferees) are moneyed and influential and often the peasants have to go to them in distress. The rural, agrarian class, among whom land alienation is rampant, are generally poor, illiterate having no knowledge about the intricacies of litigation. Going to the court to seek redressal is beyond their capacity. The process of litigation is extremely time-consuming and expensive too, which renders it more difficult for the peasant to fight against strong opponents. Therefore, they remain as before, in the clutches of the money-lenders or rich landowners with their debt burden ever increasing, as there are no particular rules regarding 'sale' or 'purchase' among the indigenous people.

The foregoing analysis brings out certain bare facts regarding rampant land alienation occurring in the state. It can be well understood that this phenomena cannot be put off by mere legislations. The obvious factor for land alienation here is perpetual indebtedness of the peasant community. A frontal attack has to be made on this problem. The big landowners wield social and economic power in the region, they retain a giant portion of cultivable land through dubious means of 'benami' and other similar procedures and they will not change their nature. Those small and marginal farmers who are prone to land alienation cannot be saved through external forces like acts, legislations and policies. Hence, a constant review of available arable land, organising the farmers prone to alienation and a constant vigil on the dismal economic condition of these peasants is needed. Small, fragmented holdings though numerically more do not exactly help agricultural productivity, such farmers often 'lease' out their land to bigger landowners who may put it to perfect utilisation. They can provide credit, fertilisers, seeds and also irrigation and enhance productivity and thus increase revenue returns too. Distribution of surplus land procured from ceiling imposition cannot solve the crisis resulting from land alienation since the accompanying necessities (seeds, fertilisers, technology, irrigation etc.) do not come with it. With time, peasants, irrespective of their race, clan or ethnicity, have developed an intricate social-economic-psychological relationship with their piece of land, and do not want to lose it at any cost. But this phenomena cannot be checked as governmental apathy, negligence, not giving equal weightage to both the segments of the population simultaneously acts as a barrier. It is quite obvious that neither of the group is well off so depriving one at the cost of another will only instigate hostility giving occassional vent to their pent-up desires and resulting in bloody riots. The officers, village-level workers need to evolve a more humanistic approach towards these farmers and pay heed to such relationships. Ignorance of market viability leads the peasants to produce for consumption only which ties him up to a hand-to-mouth situation throughout. Here again if the government could come forward with training programmes and taught the peasants about congeniality of soil, climate; supplied them with techniques seeds etc. at subsidised rates; offered them sound credit facilities the vulnerable peasants would not have to go for these things to the unscrupulous money-lenders. Years of such bondage have impaired their economic freedom, so a sense of neutral external help may be able to pull out these heavily indebted peasants from the clutches of village land-owners. Moreover legislations are made to be adhered to and not to be overseen. Pamphlets, lectures made in local dialects can make the target groups conscious about the provisions of the laws. The practice of paying cash compensation does not seem to help the peasants actually and so should be done away with, project authorities be empowered to implement rehabilitation scheme prior to the starting of the project itself. Even if cash compensations are made, it should be given at a time and within a month of acquisition of the land, and in these cases the acquisition should constitute only a portion of the landholding. However, landlessness as is understood cannot be totally uprooted, so in that case, able bodied men and women may be given employment (either in agriculture or other projects) as unskilled or semi-skilled labourers according to his/her aptitude. These measures might work in a positive direction to cure the disease called land alienation, and remove the depression caused in the economy.

NOTES AND REFERENCES

1. Agarwal, A.K. : *Economic Problems and Planning in North-East India,* Sterling, New Delhi (1987), pp. 344-346.
2. In earlier days the contradiction between land owners and their tenants were not so sharp, hence tenancy acts were only initiated in the 1930s.
3. Permanently settled areas were acquired in 1950 by the Assam State Acquisition Act, 1951.
4. Roy Burman, B.K. : *Demographic and Socio-Economic Profiles of the Hill Areas of the North-east India,* New Delhi, 1970, p. 193.
5. Goswami, P.C. and Saikia P.D. : *Problems of Agricultural Development in Tribal Areas, Indian Journal of Agricultural Economics,* Vol. XXV, 1970 (Research Report), p. 143.
6. *Report on Socio-Economic Review,* 1979-80 : Bureau of Economics and Statistics, Mizoram Administration, 1981, p. 32.

7. *Pemberton's Report* : Appendix III, 1835, A Gazette Notification (TSA).
8. A Gazette Notification (date could not be traced) regarding the dispute about eastern boundary of the Tripura State (TSA).
9. A letter of the Minister U.K. Das dtd. 11 Oct., 1891 (TSA).
10. F. No. 40 (1) Rev/58 - Record of the State Archives (TSA).
11. Das, J.N. (ed.) : *A Study of the Land System of Tripura,* Gauhati. Law Research Institute (Eastern Region) Vol. III, 1990, pp. 19-23.
12. Quoted from the Proceedings of Shashan Sammelani (Administrative Reports), 1323 T.E. (1913-14 A.D.) p. 7.
13. *Law of Tenants and Landlords—On the relationship between the landlords* (Talukdars) and tenants (1886).
14. F.No. 73/89/59 Parliament. A memo submitted by the Tripura State Kishan Sabha, 1954 (TSA).
15. F.No. 73 (32) Rev/58 - found at the State Archives.
16. *Tripura Administrative Revenue Department* (1882-83) : Act VII of 1885-Rev/69-152/57 dtd. 17 March, 1958 (TSA).
17. *Ibid.*
18. Dutta, B.C. : *Udaipur Bibaran* (Thirty years in Tripura), Agartala (1340 T.E.) p. 66 Translated from Bengali.

    The Magistrate-cum-Collector in the king's government gave an account of Udaipur division (1930)–About 5783 ha. of land is cultivated under rayati rights and khas lands and a total land revenue of Rs. 36.664 is yielded. The 'Jotes' are not very large and the plots are also small-sized.
19. Hunter, W.W. : *A Statistical Account of Bengal,* Vol. VI, London (1876), p. 503.
20. F. No. 4138/XXI/I (A/DM/SC/58) - TSA (unpublished).
21. Case of N.G. Das, F.No. 11-109/Acq/Rev/52-59 - TSA (unpublished).
22. *Land Acquisition Department*–No. 2509-10/DM/LA/SS/60/54 dtd. 4 March. 1957. (TSA) (Unpublished).
23. Sen. Tripur Chandra : *Tripura in Transition* (published by the author), Agartala (1970), p. 6.
24. *A Handbook on Cadastral Survey*—Land records, Land Reforms and Maintenance of Records in Tripura, Directorate of Land Reforms and Settlement, Government of Tripura, Agartala, pp. 2-3.
25. *Annual Administration Report of the Political Agency, Hill Tipperah* (1878-79) : Prepared by Uma Kanta Das, Assistant Political Agent. (Ex-Officio Political Agent–F. Jones) submitted on June 26, 1879 in D.K. Chowdhuri (ed.) Administration Report of the Political Agency. Hill Tipperah (1878-

79, 1889-90) Vol. II, Tripura State Tribal Cultural Research Institute and Museum, Government of Tripura (1996) Editorial Note (pp. XV-XVI).

26. Report of the Bengal Provincial Banking Enquiry Committee under the Chairmanship of K.C. De (1929-30), Vol. 1, p. 74.
27. Annual Administration Reports : *op.cit.*, p. vi.
28. Chatterjee, Partha : *The Land Question of Bengal,* 1920-1947, Vol. I, K.P. Bagchi & Co., Calcutta, 1984, pp. 202-203.
29. Tripura Gazette vide No. 3763-64/25-12, dtd. 7-4-52 and 14098 99 R/ 25-12 dtd. 2.12.52.
30. Tripura Gazette Notification No. 3111-49/DM/LA/SN/119/54-TSA.
31. Bhattacharjee, S.R. : *Tribal Insurgency in Tripura—A Study in Exploration of Causes.* Inter India Pub., New Delhi, (1989), p. 139.
32. *Tripura Land Revenue and Land Reform Act of 1960,* No. 43 of 1960. Government of Tripura.
33. Handbook on Cadastral Survey : *op.cit.,* p. 5.
34. *Ibid.*
35. *Amendment of the Tripura Land Revenue and Land Reform Act,* (1960), 1974. Tripura Act, No. 7 of 1974. Government of Tripura.
36. Agricultural Census Reports, 1970-71, 1980-81, Directorate of Statistics, Agartala.
37. Figures were taken from the Land Revenue Department of the Government of Tripura the office of the District Magistrate, Agartala (1977, 1985, 1998).
38. *Ibid.* (1986).
39. Statement showing Allotment given in West Tripura District up to April, 1998. Office of the District Magistrate, Government of Tripura, Agartala. (Some statistical evidence given regarding allotment of land to (a) landless and homeless (b) landless and (c) homeless families since its inception.

9

Observations and the Need of the Hour

It may be recalled that the objective of this study is to put forth the various manifestations of the land systems of the previous and the present century and assess its impact on land in the face of the huge population increase. Hence the foregoing analysis depicts the various causes which has led to a general rural-land availability crisis. Obvious geographical factors such as land loss due to natural reasons, hilly terrains, extensive forests combining with certain unavoidable factors such as diminishing of boundary, partition, creations of reserves, autonomous councils, reserving of forests etc. are responsible for the present crisis. Analysis of rural land-use cannot be set aside from the wider societal context. It may be remembered that in Tripura agriculture often plays a crucial role in developmental planning as the bulk of its population is employed in this sector. Industrial development cannot gain a stronghold due to several infrastructural bottlenecks which could have relieved the pressure on land to a considerable extent.

It is observed from the foregoing discussion that rural land availability did not pose any constraints even in the last quarter of the 19th century. But as population grew land availability began to get scarce. There were two types of cultivation in the state which blended quite well with each other, but the primitive mode of cultivation i.e. shifting added to a colossal loss of land as the top soil lost its nutrients due to its unscientific method. Moreover, the topography being undulating plain land needed for settled cultivation did not suffice. Terraced cultivation could have served the purpose of increasing crops production to some extent, but here hill soil being of laterite variety and not very rich in minerals could not be optimally used. River basins are of the alluvial type so are very fertile, hence the population try to concentrate in these areas lowering the man-land ratio considerably. The already inflated population cannot be decreased undoubtedly hence land classification according to its capability is to be undertaken. Land is to be provided for a variety of reasons, development, urbanisation, security projects, habitation, forests, hydel projects, in short each activity of mankind is related to land requirement other than the necessity of food products which is continuously on the rise keeping pace with the ever increasing population of the state, here optimum use of land will only be able to solve this problem as also stop wastage of land in the face of crisis. Human beings can be considered both assets or liabilities in a society according to their contribution. In the foregoing analysis, the population has been portrayed as a liability because of the mass scale immigration pattern and there being no source of sustenance other than land. This unprecedented rise in the state population without a parallel rise in their economic pursuits affected the state socially and politically too and may be attributed to be the root cause of the ongoing militancy in the state. If this huge population can be channelised in a profitable way it may earn enough returns for the economy, within the limited resources available.

The land question of the state, therefore, puts forth the basic issue of providing land as well as catering to the needs of the rising population, in short a balance needs to be maintained between the two factors. The state governments have undoubtedly tried to keep this in mind while sketching out its programmes. Shifting cultivation, as seen in the analysis is a basic aspect of the tribal way of life but this ran into rough weather as the state governments have a general tendency to derecognise this rudimentary form of cultivation which led to a crisis laden transitionary phase, putting the tribal polity in a turmoil manifesting in a vicious circle as well as creating hierarchy in a classless society. Undoubtedly, these shifting cultivators are aware of the changes taking place but the pace at which they have been subjected to change has only deepened the crisis. They are willing to change over but land cannot be provided abundantly so optimum utilization of land remains a question yet to be answered. It is not that land availability is scarce in all parts of the state. As for instance, border lands are still lying vacant in most regions, fertile tracts fit for crop production extensively, but owing to a deteriorating socio-political relation between the two neighbouring countries, miscreants are active in these areas. Consequently though fertile land is abundantly available, they lie abandoned. The owners have shifted to nearby towns taking up some other economic pursuit and even those without land are not quite eager to tread in these manforsaken regions. Forests are an absolute necessity in today's environmental condition as it maintains the ecological balance so necessary for survival. But the dilemma whether to preserve forests or rehabilitate the people could not be answered and the government seem to do both simultaneously i.e conserve forests pertaining to the various rules and Acts and at the same time open up forests for developmental needs, projects, security encampments etc. This does not solve the land scarcity problem rather leaves one stranded as both the greenery as well as rehabilitation is required. It is seen, that the traditional custom-bound simple autochthons' plight seemed more formidable in the face of change. Parochial sentiments, antagonistic attitudes among them led to granting of autonomy to these autochthons in order to protect their values as they wanted. But it has come to be noticed that even such autonomous bodies cannot cater to all the basic demands of the entire tribal people (only those who reside within the reserves receive the privileges). Hence this step too has not been able to solve the ethnic racial strifes and the failures and their causes analysed show faulty planning and too much dependence upon the state government which delays policy implementations occassionally. The legacy of the by-gone century's land systems is most openly manifested in the alienation of land, both cultivable and habitation, mostly seen among the rural masses and which is perhaps the raging problem at present in Tripura's context. The cause dealt here clearly indicates how fertile, irrigated land concentrated in the hands of the top privileged class who were often absentee land-lords. Such settlements added to a steady increase in the landless class too. Even the introduction of the Tripura Land Revenue and Land Reform Act of 1960 or its later amendments failed to check the transfers as loopholes were only too obvious. Although the primary purpose of the said Act was to prohibit transfer and alienation of land and help to plug the loopholes but it could hardly prove effective. This Act, aimed merely at safeguarding tribal land irrespective of the status of the transferee. So the unavoidable happened, land was transferred from poor to rich tribal landowners without any prohibition or from tribal peasant land owner to settled peasant landowner under covers as the ceiling limits were quite high. Hence land alienation came to be viewed as a symptom of general depression in the economy as mentioned in the analysis.

Undoubtedly, the governments of the state has stepped in with numerous measures to solve the economic crisis of the state arising out of rural land scarcity, but certain anomalies in policy implementation leave gaping holes. For one, it may be stated, that re-distribution of land has provided land to many landless peasants, yet this has not been able to significantly reduce the inequalities in land distribution. Till now the mass peasantry comprises of submarginal, marginal and small farmers whose economic viability is doubtful. Such farmers remain at a low level of subsistence and are led into debt frequently and since land transfers (tribal to non-tribal) are disallowed, owning of land remains out of question for quite a sizeable peasant population even with governmental doles. Restored land is often not acceptable as they shy off from repossessing such land fearing the wrath of the village money-lenders or traders. Moreover, access to non-agricultural work is limited and seasonal too. Being unskilled, this sector remains closed to the peasants, hence the rise in the landless rural population. Whatever developments have taken place remains in the clutches of large landowners. Even if certain monetary benefits percolate to the lower rungs they use it up in no time in merry-making, spending on consumer durables, hence no marked impact of governmental schemes and programmes can be observed. Land availability, as is understood is scarce and so introducing any developmental scheme for the peasants of the state is a challenging job. One must remember that two segments of population are thriving in the state and similar treatments is to be meted out to both in order to avoid creating any schism in their minds. Dealing with the indigenous peasant population one cannot ignore the presence of shifting cultivators in large numbers. Uprooting them all of a sudden from their environment could be disastrous so gradual moves were being made e.g. terracing the undulating terrains through subsidy schemes. But these cultivators could not take advantage of the same. There was a growing feeling amongst them that development would lead to an erosion of their traditional rights over land resources. It is not that they are ignorant of the modern techniques of agriculture but their basic nature of keeping away from mainstream hinders them from taking advantage of the government subsidies and doles. Rather a sense of antagonism was created in their minds as they felt that they were outcasts in their own land since forests were reserved, their daily requirement of fuelwood timber were sidetracked; and this feeling strengthened when they saw that the Reserve forests were opened up to accommodate the displaced immigrants from erstwhile East Pakistan. The hapless shifting cultivators decided to enter the colonies having no other alternative at the moment. These colonies were set up in the outskirts of the towns, but lack of enough arable land, subjected to a new technique of cultivation and ignorance restrained them from embracing this new lifestyle and hundreds deserted the colonies in search for livelihood elsewhere (working as daily rated unskilled workers in urban areas, coolies or even rickshaw pullers).

Thus, if the problem of rural land scarcity is to be tackled effectively, substantial, improvement of land distribution schemes, crop production skills, land capability as well as employment opportunities in the agrarian sector needs to be harped at, whatever the outcome might be, the rehabilitation schemes should not be abandoned as it is the only answer to bring these shifting cultivators on the same plane with the mainstream population. Land available being subjected to continuous shifting mode of cultivation since long is gradually depleting the soil cover turning it unproductive which cannot be accepted any more. The need now is imperative, alternative to fit the growing aspirations of the evolving society into the traditional value system of the state. Material development at the cost of hold values of life and cultural heritage lead to an unhinged society disturbing the social equilibrium. Legislative policies, enactments have been made since Independence with the aim of solving the scarcity question, yet the present situation shows no remarkable achievement in this field. The Tripura Land Revenue and Land Reform (1960) Act was passed but this too was plagued with several anomalies as the provisions were essentially directed towards a particular segment of the population. This obviously leads to land use conflict in the rural sector and any physical manifestation of the problem reflect more fundamental conflicts involving the traditional values and socio-economic organisation of the society. Hence it is quite clear that such one-sided movements cannot help in lifting the state from abject land scarcity crisis.

This fact is more understandable when it is found that the autochthons are not the sole sufferers—the plight of displaced plainlanders who came from the neighbouring East Pakistan puts forth a similar if not identical picture. One may be reminded that not all were actually "invited", scores came on their own either lured by better prospects of living or to escape socio-religious atrocities, hence not all were lucky enough to receive government dole of land and monetary benefit to settle. They started to share, mortgage or transfer land even encroach if need arose which made land availability shrink further. Such process evicted many former tenants from their tenancy of new tenants were ready to pay more. As land got reserved for several other reasons viz., forests, autonomous councils, developmental projects, roads etc., these new tenants faced severe land availability crisis adding to their woe. A peculiar consequence can be noticed here, a struggle for control of land is being waged between the plainlanders and the indigenous people though after all these years both can be termed as locals in the strict sense of the term. Legislation and enactments have been passed to protect the peasants but the present state of affairs which are mostly historically evolved do not bear testimony to the same. The fickle policies of the government, opening up of tribal reserves to accommodate the displaced immigrants and then once again setting up council boundaries, created much animosity in the minds of the users of land who were unaware of their land ownership rights. Forcefully taking away of land by one segment antagonised the other and this resulted in the outbreak of violent outbursts during the 1980s. The 1990s witnessed rise of militancy and the hapless citizens are at the receiving end. Failure on the part of the government to implement policies properly may be a reason for such outbreaks. One should not forget the incomparable development potential of the land, the mistakes of the yester-years can be rectified perfectly. Thus the need of the hour may be discussed here as the study has observed in the foregoing analysis the evil impact that land scarcity is having on the social and economic plane. Catering to this, following suggestions will perhaps help in organising the agrarian sector, pacing up production as well as pacifying both the segments of the population.

The first and foremost problem to be dealt with is the ongoing militancy in the state. Though at one point of time it was felt that granting of autonomy could solve a number of problems relating to the autochthons but the state has witnessed a plethora of insurgent groups, who strike time and again causing much loss to life and property. Those involved should be made to understand that insurgency is no way to get back their long lost land, and other rights. Though this is easier said than done, a beginning may be made by initiating dialogues with the militants, setting up effective intelligence network so as to comb out hardliners, discussions should be made on equal lines and not hover around the demands or luring the militants solely. Undoubtedly wooing the insurgents with jobs is an attractive proposal and a tremendous psychological trump card too, but then such jobs are inadequate to accommodate all and often given to unsuitable candidates who are not satisfied with these jobs. Moreover, eagerness of these new employees to spend on consumer durables siphon off their earnings and they once again revert to militancy putting the problem back to the square one. A two stage action plan needs to be warranted - a combined all out counter insurgency operation launched by the security froces with the aim of destroying the insurgency permanently. The farcical "misguided youth" syndrome needs to be done away with. Along with this a massive economic renaissance with an aim towards rapid development undertaken. Perhaps talking them into surrendering, instilling in them the felling that the plainlaners are not essentially usurpers, rather, they can together build up the economy may attempt at solving the problem. Yet, it needs to be pointed out here that such a herculean job cannot be that of governmental forces alone and so non-governmental organisation social workers can be roped in for this purpose to gain the confidence of the waylaid people. Mass propaganda against such activities aiming at earnest removal of the roots of grievances rather than insurgency will be the right track if special care is taken to look into the socio-economic security of these frustrated people, so that they do not resort to infighting even.

Once the insurgencies are combated the next question that arises is where and how to settle those without land. Jobs according to capability seems to be the best answer, but a majority of such population is of cultivation stock and so are more familiar with land-related jobs only. Perhaps, plantations are best suited viz, tea, coffee and ruber as these do not require mush skill or expertise and can do with little training only. These plantations are not only congenial to the climate but being undertaken extensively can provide job to the labourers throughout the year. Hence, it will fetch more income for the state, while helping the landless to tide over their economic hardships. Moderate gradients, even degraded ones need to be brought under cultivation by applying technical know-how since much irrigation facility is not needed here. The Tripura Rubber Plantation Corporation Ltd., (1983) has already made a beginning in this direction.[2] A landless Jhumia is allotted 1.5 ha. of khas land and 1.6 ha. for homestead and other purposes. A total of 1427 families have been settled within a span of 6 years (1984-1990) 2140.5 ha. of land.[3] However, this is not enough and the scope for provision needs to be widened. Simultaneously soil needs to be subjected to appropriate treatment so as to augment the benefits from land as well as improving the environment. With growing economic development of the state the timber requirements too go up but most of the valuable trees come under the "reserved" status. Raising such trees in waste land, beside catchment areas can provide with much financial help to those involved as well as help the soil to recover its productive value too and establish a symbiotic relationship between the forest and the dwellers. Some experts are even of the opinion that such plantations will minimise flood intensities as soil erosion can be done away with. However, the age-old practise of shifting cannot be done away within a matter of years. Patient outlook, preparing a base for the modern technology and helping those involved to wean over to other land related work may be a step forward.

Imposing new, alien methods should be done cautiously or otherwise it will only add fuel to the already raging fire and ultimately lead to disintegration of the society, while individuals may gain a stronghold which is not quite desirable. Therefore, colonies are to be set up in such regions which are familiar to them, such as, the hillocks, the gentle slopes, along the rivulets of streams, which would mean a literary rehabilitation and would not generate any feeling of alienation or that of being uprooted from their original dwellings. Such socio-economic environment will give the cultivators a sense of security and solidarity whereas a forceable removal would create antagonism in their minds facing complete estrangement from their age-old customs, traditions and familiar surroundings. Perhaps, in such an atmosphere, jhuming will not at all be degrading to the soil and environment as also subsidiary occupations like poultry farming, animal husbandry, silk-worm farming be introduced which would lessen the pressure on land. And even if land is restored, it ought to be clearly demarcated, reclaimed and developed before allocation. Land records are something unheard of by the inhabitants, therefore, in order to annihilate any spurious land documents a thorough perusal of deeds, registrations need to be done. If their land owing rights are honoured, which is their basic demand, the developmental schemes can be more adequately undertaken. In fact, field studies indicate that given help and landuse knowledge, the degraded regions can attain, through their hardwork, significant achievements within a short span. New occupation can be resorted to which are nearest to their way of life and traditions, by utilising the abundantly available labour, this would undoubtedly relieve the pressure on land to a considerable extent and that too by voluntary will of the autochthons. In doing so, the shifting cultivator will be able to overcome the deprivation he has faced all along as well as the resultant drudgery of landlessness and joblessness during the transitionary phase. In due course they will forget that shifting was a way of life for them and gradually become proud owners of plantations, orchards, subsidiary industries and so on.

Afforestation policies are being undertaken on a large scale. At the moment, forests contribute to 62 per cent of area of the state. The study has observed that though forestry is spread over such an extensive area, encroachments are rampant too which does not help to retain the greenery. If such legislation could be passed wherein forests were 'opened up' i.e., the products used by those inhabiting them or living on the fringes, perhaps encroachments and consequent destruction could be minimised to a considerable extent. Allowing such dwellers to extract limited forest products without harming or destroying the environment (e.g. collection of fire-wood, dry leaves, fruits or even for grazing) encouraging them to plant saplings, engaging them in de-weedification process may strike a balance between the land use systems of the peasantry and the conservationist policies of the state. Modern concept of forestry views it as a multi-purpose, socio-economic activity, which means application of technical methods, by raising plantations of commercial species. Forcefully restraining the people from enjoying its products often lead to grave social disharmony (smuggling, rioting, non-observance of rules etc.). Therefore, the fringe-dweller's fuel and fodder requirement may be met amicably, and pursuance of such a standpoint would deter the dwellers from flouting forest regulations and at the same time help in maintaining and preserving them. An example from a neighbouring state, Nagaland, may be cited here.[4] At Khonoma, the tribals 'construct contour bunds in jhum fields.... and grow alder plants which they take care not to destroy even during Jhum. As a result, within three years of jhum cultivation, the entire land is covered by green foliage'. That means, though following shifting mode of cultivation the cultivators take care of the forestry conservation methods. If such a scheme can be followed in Tripura it would be able to solve the dilemma of whether to conserve forests or open up land for cultivation. There is only 0.1 per cent of village/community/unclassified forest in Tripura, where as there are 65.3 per cent reserved forest area with a potential of 34.6 per cent where no tribal or non-tribal forest dweller is allowed to tread.[5] Right training and education can give such people the knowledge of forest conservation, on which they have to depend and if understood, they would surely not like to destroy the same. The term 'Participative Management' can be introduced here which will help forestry play a significant role in development through an adjustment of land, capacity, peoples' aspirations as well as project requirements. Reservation of forests in degraded waste-lands such as catchment areas of hydro-electric projects, abandoned jhum land will be helpful in improving the ecological balance substantially. Those who have to use the forest products are not the owners of the same, hence they do not bother to protect the resource too. This is an inequitous condition and needs to be changed wherein the user should be accepted as the management partners to develop a healthy working environment and dispense with the idea of a casual labour force, which benefits none. Agriculture and forestry need not be treated as two mutually exclusive arenas of development but can be made interdependent, as both require a large labour force which is amply available in this state. Those who are not particularly interested in settled cultivation might be absorbed in this sector alongwith the ever increasing migrant labour force the number of which often disrupts forest harmony besides changing the demographic structure of the adjoining area. Any contradictions resulting from the pressure on land may be avoided by following an integrated approach, of course, for this the 'Reserved' or 'Protected' forests concept has also to change. Those who are using forest produces are now being inducted in the management of the same hence he becomes a grower in due course, this will eliminate the middle-men, the encroachers, thus helping harmonising the interests. In such an atmosphere, expert guidance can also lead to growing of economic species in the depleted forest areas with limited rights in favour of the individual. Again, it is quite obvious that only government initiative regarding such issues cannot be enough, unless social workers and individual too participate thus contributing in the process of decreasing the pressure on land in Tripura.

Land ownership rights did not get much preference in the planners proposals till date owing to a defective land system prevalent in this part of the country where in actual tillers are not necessarily the owners. As immigrants came in, a large number of landless labourers came into existence. Reports give evidence that both Assam and Tripura lie very close to the bottom of productivity ladder among major states of India, which prove their low land availability. The Land Reform Acts have effectively abolished landlordism, tenancy and sub-tenancy in agricultural sector but a single class of self operating peasants could not be created yet. The hills and hillocks have not been cadastrally surveyed neither have there been any hydrological surveys. Hence people still depend upon customary laws for discerning their land-ownership rights, which obstructs any technological input for raising production. The need of the hour, is therefore, preparing record of rights for the users as soon as possible. It should be kept in mind that the approach should be humanistic along with a legal attitude to do away with any ill-feelings amongst the native dwellers. The observation of Land Reform Commission for Khasi Hills, (1978)[6] may be recalled here, 'We cannot talk of a development without a plan for no planner can properly prepare any plan for the development of agriculture and industry based on it unless he has statistics about land and its use. Therefore, it is quite obvious that recognition of individual rights over a piece of land will encourage the owner to adopt measures for permanent improvement of the land as now technological financial aid will flow in smoothly. This in turn will lead to improved and increased production, a result of the sentimental attachment between the user and his land. Simply fixing up of ceilings, restoring lands granting khas land will not help in solving land scarcity.

The most unpleasant outcome of all these problems is the 'land alienation' factor which can be hardly put to an end in the face of increasing population of the state, even by the legal actions taken against it. Certain discrepancies, loopholes are always resorted to in these acts and policies which enable the mighty to hoodwink the poor peasants and encroach upon their lands. The Tripura Land Revenue and Land Reform Act, (1960) attempted at doing so but hardly met with any remarkable success, its amendment (1974) disallowed transfer or sale of land from tribal to non-tribal with out the authority's permission but all these aimed at protecting only one segment of the population. There is no concrete rule to stop land alienation among the settlers. It is also observed that the indigeneous people often sell their land to their own people creating a new stratum of landed gentry, unheard of in their classless society. At the same time an army of landless tribal peasantry is created some of whom often get involved in insurgent activities in want of a livelihood. Mere laws, rules, acts, government policies do not seem to be the answer to stop alienating land. Land reforms are to be made in a much wider purview, not limiting it to simply restoration of land only which is certainly not quite acceptable to the transferers even who do not want to break the bond of trust and confidence with the transferrees. The actual reason may be fear of reprisals as also ethics or both. Poverty and the ensuing indebtedness are observed to be the main causes of land alienation and eradicating this is a colossal and near impossible task. Land is the agriculturist's principal asset and thus helping him to retain the same should be given priority by the government. The state cooperatives, financial bodies, government outlets need to come forward to purchase the land when the owner is in distress and return the same when the latter is in a position to buy back his land repaying the amount. Most often it is observed that these peasants are not in a position to carry on litigation in case of disputed ownership rights. The government here should provide full legal assistance and even financial assistance to continue with the same. The poverty-stricken seasonal peasants in the remote areas are hardly aware of laws prohibiting sale/transfer of land. One can correlate this fact to mass scale illiteracy among the people, for this mass campaigning need to be undertaken, even communication methods and skills need to be emphasised by using local dialects and initiating dialogues with them on the same plane which may make the masses conscious of the provisions of the acts. Land alienation resulting due to the installation of the Dumbur Hydro-electric project has come under much criticism. The sufferings of the affected people may be minimised by incorporating suitable planning for their rehabilitation by providing alternative land for home-stead, cultivation rather than making cash grants or compensations which are used up in no time. However, cash compensation may be paid to those peasants who had to part with a small portion of their holdings for governmental requirements, which do not substantially affect the economic conditions of the family. Offering employment opportunities in the projects which displace the people may also be undertaken so that the peasants do not get indebted leading to a chain reaction ultimately resulting in land alienation. But they should not be exploited by the middle-men and paid low wages, which again will lead to accumulation of total distrust for contemporary government and give vent to insurgency.

Another indirect consequence of land scarcity problem manifests itself in the absence of any surplus production to be sold elsewhere. As in the past, the productivity still hovers around self consumption leaving little surplus. Certain essential inputs e.g., fertilisers, pesticides, high yielding variety seeds, along with expertise may lead to an increased production even in small holdings. However, it may be pointed here that not all holdings need produce the staple food for the society, emphasis need to be laid on locality development as well as developing local skills and resources. Fruits like pine-apple, jackfruits, oranges can be grown in abundance to process for pulp, juices or even beverages, citronella herb, quite abundant here is used in pharmaceuticak, may be grown commercially, mushrooms are gaining importance in the layman's diet even and this will fetch quite a lumpsum for the agriculturists; coconut industries are known for their profit oriented products which are quite congenial to the soil of Tripura manufacturing coir mats, ropes, carpets, rubberised coir shell powder, oil or even food products will be quite lucrative too; growing pepper, coffee, cashew-nuts will engage a sizeable labour force; and one should not forget the various uses of bamboo which are abundant in Tripura, if these are properly cultivated match, paper, joss-sticks industries will rejuvenate the economy undoubtedly. Jute growing will be helpful too as jute sticks can be potentially used as a basic raw materials for a number of ancillary industries like viscose rayon, paper, chip-boards, chemicals (viz., furfural powder, carboxyl-methyl cellutose, alcohol, activated carbon etc.). All such projects naturally need support from the state authority. Research-oriented studies, undertaken by the state agricultural department can do wonders in solving the emerging crises of land scarcity.

Yet, it should be borne in mind that simply recasting the agricultural sector and bringing prosperity to rural masses is not the only answer in solving land crisis. As such agriculture cannot be relied upon solely due to its incapability to changes, it is a stagnant factor as its productivity cannot increase without an increase in its technological inputs. Perhaps, during this period, rural industrialisation can step in, which if gradually introduced phase by phase stressing on agro-products will provide an apt answer. Here too one is reminded of the very first hindrance, that of transport. This aspect seems to be the greatest bottleneck in Tripura's industrialisation process. Partition of India made Tripura a landlocked territory and posed a serious hindrance to its communication with the other parts of the country. The natural hinterlands in Bengal, and its modes like Comilla, Feni, Narayanganj were taken away which used to provide easy access to the epicentre of trade i.e. Calcutta. Suddenly the ports became foreign and thus inaccessible or in other words, Tripura became one of the least accessible frontier regions of India. The terrain of the region being difficult too, surface communication posed great hazards. The plan outlay for road building was only 12.16 per cent where as in Arunachal Pradesh it stated to be 31.62 per cent. Manipur 15.12 per cent, Meghalaya 21.17 per cent, Mizoram 24.25 per cent & Nagaland 23 per cent.[7] Moreover, weak bridge upon rivulets, streams raise problems. It is quite obvious that without such basic infrastructure industrialisation will be a far cry for the state. Industrialisation also brings along several hazards like organizational, administrative etc. Those involved in such renovation process are to see that the essential social fabric of the native traditional society remains intact. At the same time local inhabitants of both the groups need to be integrated to form a social unit, so that either do not feel secluded. Such externally induced changes need to be carefully assessed to enable future course of development. Policies of development are undoubtedly sound but that should not upset the harmony that exists in the society. A smooth transition towards accommodation at new levels is required. A step towards modernisation of agriculture can be taken by initiating foreign investments or even domestic investments, a stage which may be coordinated by the North-estern Council. It is but natural that surplus labours if absorbed in agricultural sector will mean gradual but sure depletion of resources. Therefore, policy implementation should aim at both the segments of population without isolating a single one–as that leads to a demoralising effect on the population resulting in occassional communal frenzy.

Land question is the basic issue of the economy of the state and the causes for its scarce availability has been dealt in an historical perspective in due course of the study. To tackle the said problem, the need of the hour is mass contact, general awareness among the growing population and a positive outlook towards alleviating poverty and thus establishing an informal relationship. Offering jobs are undoubtedly lucrative, but at-times the peasants are unskilled for the same and a general tendency of dodging such alien jobs can be noticed. Spreading education, vocational training opportunities, offering incentives may help in bringing the shy peasants out of their cocoons. Interest free loans instead of cash grants would help the users of land to get liberated from the clutches of unscrupulous money-lenders. The grievances have been allowed to fester for long which reflect the people's angst. Patronage, even if given should be equal to both the segments of the population, lest one should feel deprived and resort to militancy. Counter-insurgency cannot solve the ongoing crisis, rather granting basic socio-economic needs would perhaps take a step forward in answering the land question of the state.

NOTES AND REFERENCES

1. Deb Barma, S.B.K. : *A Study of Jhumiya Rehabilitation*, Directorate of Research, p. 15. Bench Terracing @ Rs. 600/- per acre of land; Bunding @ Rs. 150/- per acre; Lunga reclaimation @ Rs. 250/- per acre.
2. Annual Report of Land Utilisation Projects of Tripura Rubber Plantation Corporation Ltd., (1983).
3. *Ibid.*
4. Report of the Committee on Forest and Tribals, New Delhi, 1982.
5. Directorate of Economics and Statistics, Ministry of Agriculture, Government of Tripura.
6. Report of the Land Reforms Commission for Khasi Hills, 1975.
7. Basic Statistics of North-East, North-Eastern Council, Shillong, 1995.

Bibliography

Files and Letters

Eastern States Agency No. W. 3-77/42-4571 Letter to the Chief Minister of Tripura State from Capt. A.C.K. Maunsell, dated 30-4-1943. Secretariat Archives, Agartala.

F.N.H. 28/F Letter No. C.No. XI/Con/29/Cus/49/6922. dtd.-Agartala 29.11.49. Confidential letter of the superkising Inspector, Land Customs, to the Collector of Land Customs, Tripura. Copy of the Home Secretary, Delhi. Under the Chief Commissioner's Office No. 1578/428/F, dtd. 6.12.49 (unpublished).

File No. 11-109/Acq/Rev/52-59. Regarding purchase of Jote land (Jote No. 234) by a displaced person from the Rehabilitation Department. Secretariat Archives, Agartala.

File No. 12939/DM/Rev/XI-3/59 dtd. 11.11.1959, question regarding a decline noticed in the Ziratia operated holdings; Secretariat Archives, Agartala.

File No. 1601-07/F. 1 (58)-DLRS/LR/78 General Order No. 6, dated 9th Feb., 1979, deleting the word Ziratia tenants and changing it to Rayati. Directorate of Land Records and Settlement, Agartala.

File No. 1-E/1 1 No. 580 A, Calcutta, 8th August, 1891 on the question of disallowing the Maharaja of Hill Tipperah to remove the northern embankment of Gumti by the Board of Revenue, West Bengal State Archives, Calcutta.

File No. 21746/DM/FP/XVI-9/88 dtd. 8.12.58 Prohibitory orders of the Sub-divisional officer on peasants harvesting land belonging to Ziratias, Secretariat Archives, Agartala.

File No. 2509-10/DM/LA/SS/60/54, dense jungly land at Hatileta (12.80 acres) given for refugee rehabilitation, office of the District Magistrate, found at Secretariat Archives, Agartala.

File No. 28-S/1 of 1898, 28-S/1 of 1903, 28-S/4 Mar. 1904 and 165/22 of 1904-05. Survey and Settlement of Tippera, West Bengal State Archives.

File No. 4138/XXI/1 (A)/DM/SC/58 Regarding a Khas land at Khowai belonging to a tribal peasant given to a displaced person. Secretariat Archives, Agartala.

File No. 8250-53/F.1(58)-dLRS/Estt./78. General Order No. 3 dated 5.12.1978. Technical Recording of land deserted by Bangladesh Nationals, Directorate of Land Records and Settlement, Agartala.

File No. 8B/1 4,5,6, 8B/3, 9, 10, 1, 2 of 1898. Survey and Settlement of Hill Tippera, West Bengal State Archives, Calcutta.

File No. B 26/531/1952/FN-XX-113/52. Letter No. 64-p/XX-13/52 dtd. 5-5-1953 agreement between the District Magistrate and the Superintendent of Police of Tripura on one side and the District Magistrate of Tippera, Noakhali on the other on question visa of the Ziratias, Secretariat Archives, Agartala.

File No. B/241/517/F.N. 2 (21)-P/58, Representation of the All India Kisan Sabha, dtd. 24.10.1954, Secretariat Archives, Agartala.

File No. B-3/S-21 Annual Report on the Administration of the Political Agent of Hill Tipperah, 1884-85, 1885-86. (Published in Calcutta Gazette) 1887, Secretariat Archives, Agartala.

File No. B-4/S No. 25/1963, Letter No. 339, 11-11/53 p. The Order passed by Inter-Doninion Agreement limiting the export of paddy subject to payment of export duty. Secetariat Archives, Agartala.

File No. B-51/S-14, A-3, The question of the title of the Chief of Hill Tipperah, Raja of Maharaja, Letter No. 4339, dtd. 26th Nov., 1873-77, Secretariat Archives, Agartala.

File No. B-62/S-9, Change of name of Tripura State from “Hill Tippera”, Tripura Secretariat Archives, Agartala.

Letter No. D-514 W-3-34/43 dtd. 11.12.1943 Certain rules posed regarding land cultivated by the Ziratias.

Acts and Notifications

Calcutta Gazette, First Forest Act of British India, 1865. Act VII, Section 2 Act I, 1886, 1887.

Jarip o' Bandobast Sambandhiya Niyamabali (Survey and Settlement Rules) Tripura Code Part III, 1899. (Amended in 1913 and 1926 A.D.)

Bengal Tenancy Act VIII, 1885. Revenue Department. Rev/69-152/57 dtd. 17th Mar., 1958.

Maharaja Bir Bikram Kishore Manikya's Forest notification regarding four regional forest areas Ramkrishnapur, Chandrapur, Garjee and Kachigang No. 378/3/87 dtd. 6.1.1349 T.E., 1939. No. 7853 B/3-83 (1938); No. 4160 B/3-65 (1937); No. 35 B-3-85 (1939).

Tripura State Gazette Sankalan, Directorate of Education, Government of Tripura, No. 223 dtd., 16.6.1353 (1943).

Rajasya Sambandhiya Niyamabali, 1880 (1290 T.E.) dealt with land tax from Kayemi Taluks (i.e.; Permanently settled estates), Khas Mahal (i.e.; government owned land) and karsha praja (cultivating lenants) etc.

Tripura Tenancy Act right to acquire tenancy land (1296 T.E.) 1886 A.D. amended in (1355 T.E.) 1932 A.D. Regarding Jote or raiyati rights, but does not define the status of under-raiyats (Korfa tenants) F. 73-(32) Rev/58. Secretariat Archives, Agartala.

First Forest (Reservation) Act, 1886 of the Princely regime.

The Tripura Land Revenue and Land Reform Act, 1960 (No. 43 of 1960) Amendment in 1974 (Act No. 7, 1974), Amendment in 1975. Tripura Government Press, Agartala.

The Tripura Tribal Areas Autonomous District Council Act, 1979 (Tripura Act 12 of 1979) Law Department, Government of Tripura.

Gramya Mandali Ain (Village Mandali), 1940. Act 1 of 1350 T.E., 9th Asharh.

A Gazette Notification regarding the border between Manipur and Hill Tipperah–Pemberton's Report : Appendix III 1835. Secretariat Archives, Agartala.

Praja–Bhumyadhikari Ain, 1886–Law of Tenants and Landlords defined on the relationship between the landlords and tenants.

Gazette No. 3763-64/25-12 dated 7.4.1952 and 14098-99. R/25-12 dated 2-12-1952 for acquiring land to extend Military Cantonment, Secretariat Archives, Agartala.

Notification No. 3111-49/DM/LA/SN/119/54 - a state government notification regarding release of land taken by the Military department, Secretariat Archives, Agartala.

File No. F/2-2/5 For-80/CD, dtd. 9.8.82 Rescinding the 1952 notification of the Forest Policy. Forest Department, Government of Tripura, Agartala.

File No. 40 (1) Rev/58 and File No. 73 (32) Rev/58 - Regarding authentic documentary evidence on land records and occupancy rights of the tenants. Secretariat Archives, Agartala.

Forest Rules of Tripura, Notification No. 3, Tripura Administration, 1957. Government of Tripura.

Reports

Reports of the Political Agency, 1872 on the influx of Mogs in Tripura, West Bengal State Archives.

Bengal Administrative Reports, 1882-83, 1889-80, 1914-1915.

Tripura Gazette Extraodinary Issue, May 18, 1957, February 26, 1972.

Report of the Land Revenue Commission, Bengal, Vol. II under the Chairmanship of F.L.C. Floud, Alipore, Bengal Government Press, Government of Bengal, 1940.

Annual Report of the Archaeological Survey of India, 1925-26, 1930-34.

Unpublished Reports of the Tripura Administration (1343 T.E.).

Census Report of Bengal and Sikkim - Part I Vol V by A.E. Porter. Vol. XXVI Tripura, Part I (c), 1976.

Report of the Ministry of Food and Agriculture. National Forest Policy, Government of India, 1952.

Report of the Commissioner of Scheduled Castes and Tribes, Manager of Publications, Government of India, 1955.

Report of the Study Team on the Union Territories and North East Frontier Agency, Administrative Reforms Committee, Delhi, 1968.

Census of India 1991 series 1 paper 2, 1992. Statistics Department Government of India.

Report of the Dhobar Commission on Land Alienation of Tribals.

A Report on a comparative study of Crop Production under shifting cultivation–A case study of Garo Hills, Meghalaya, Agro-economic Research Centre, Jorhat, 1980.

Bulletin of the Land use Board on Classification on Lands data using MDAS, Agartala, Tripura, 1993.

Report on the land of the tribals, National Committee, New Delhi, 1981.

Report of the Tripura Rubber Plantation Corporation, Department of Forestry, Government of Tripura, 1983-84.

Report of Tripura Forest Development and Plantation Corporation, Department of Forestry, Government of Tripura, 1984.

Report of the Planning and Coordination Department, Draft Annual Plan, Government of Tripura, 1993-94.

Report of the Task Force on Shifting Cultivation (1983), North-Eastern Region (Census Extract), Government of Tripura, 1997.

Report on Agricultural Census, Tripura Vol. II, Agricultural Census Commissioner and Directorate of Land Records and Settlement, 1985-86.

Report of the Land Use Board Land Use and Land Cover Change. Annual Perspective Plan, 1993. Agartala, Government of Tripura (1993).

Annual Report of the Tripura Tribal Areas Autonomous District Council - Three Years of work 1982-85. The Last Ten years (1985-95).

Report on Distribution of Forests by Legal Status in the North-East, Directorate of Economics and Statistics, Ministry of Agriculture, Delhi, 1984-85.

Narrative Report of the Hill Tipperah, North Chittagong and Lushai Hills - A Topographical Survey, Capt. Badgley, 1872-73.

Report of the Agricultural Finance Corporation, A Study of Shifting Cultivation in Tripura, Agartala, 1983.

Reports of the Planning Commission Plan Outlay (Five Year Plans) Sixth Plan, Government of India.

Report on Socio-economic Review, Bureau of Economics and Statistics, Mizoram Administration, 1981.

Report of Satellite Remote Sensing Survey of Natural Resources of Tripura, National Remote Sensing Agency, Vol. 6, Nos. 324, 1985.

Administrative Reports of the Tripura Government, 1347, 1348 and 1349 T.E. (1936, 1937, 1939 A.D.) Sections 176-178.

Annual Administrative Reports of the Political Agents, starting from 1872 onwards, West Bengal State Archives, Calcutta.

Journals, Newspapers, Seminar Papers and other published and unpublished documents

Journal of Asiatic Society of Bengal, Letters XVIII, 1951.

Journal of Asiatic Society of Bengal, Letters XIX, 1850.

Eastern Bengal District Gazetteer, Tippera (Webster J.E.) Allahabad, 1910.

Parisankhyan Barta, Vol. I, No. 2, Government of Tripura, An article by J.B. Ganguly, Pattern of Occupational Structure of the Tribal Population in Tripura.

Social Research, Vol. 3, No. 1, Jan-Mar., 1983. An article by A.K. Agarwal, Land Ownership System in the Hill Areas of the North-East Region and its Effect on Development.

Manual of Indian Forest Botany, ed. by M.N. Bor, OUP, Delhi, 1963.

Economic and Political Weekly, Vol. 21, No. 23, June 7, 1986 Article namely Influx from Bangladesh, No. Solution offing.

Economic and Political Weekly, Vol. 35, No. 39 September 28, 1990. Article by H. Bhattacharjee, Communism, Nationalism and Tribal Question in Tripura.

Frontline, September 17-30, 1988. Article by M.S. Prabhakar, 'Tripura : The Accord and After.'

Chunta Prakash (A Bengali Monthly Organ of the Rural Area of Tripura District & ed. by A.C. Bhattacharja), 16 January, 1930 (an extraordinary issue). An article by Girija Mohan Sanyal, The Cultivation in Tripura.

Proceedings of Tripura Legislative Assembly :

22.3.1966 (A Resolution by Sudhanna Deb Barma).

15.3.1967 (Amendment moved by Aghore Deb Barma).

15.12.1967 (Private Members' resolution by Aghore Deb Barma and supporting speech by Abhiram Deb Barma and B.C. Deb Barma).

Memorandum Presented to the Prime Minister, Jawahar Lal Nehru by Sri Dasarath Deb, Sep. 10, 1955 *Secretariat Archives*.

Sentinel (The), 17 Dec., 1986. An Article by N.A. Gokhale, Chakmas–A persecuted Lot.

Statesman (The), July, 1993, An Article by H. Bhattacharjee, TUJS Alienated from the Tribals. (Calcutta Edition).

Statesman (The), Editorial April 12, 1993. (Calcutta Edition).

Statesman (The), 26 Jan., 1995. An Article by S. Chakraborty, Politicians make a Sport to Themselves. (Calcutta Edition).

Eastern Bengal and Assam District Gazetteers, R.H. Sneyd Hutchinson, Allahabad, 1909.

Desher Katha (Daily), (A Bengali Daily published from Agartala), Mar., 24, 1983 an Article by J. Deb Barma, Setting up of Tribal Development Blocks.

Syandan Patrika (A Bengali daily) 12 Dec., 1991 and 20 Jan., 1992, Articles on Regional Autonomous Councils for Reangs and Mizos of Tripura.

Souvenir of Tripura Employees Coordinators Committee, Fifth Biennial Conference Feb., 13-16, 1981. Articles in Bengali, by Sri Dasarath Dev on Ganamukti Parishad's attempt at Reserving Tribal Land and Role of Ganamukti Parishad in the Democratic Movement of Tripura.

Indian Journal of Agricultural Economics, Vol. XXV, 1970. Article by P.C. Goswami and P.D. Saikaia, Problems of Agricultural Development in Tribal Areas.

Proceedings of North-East India History Association, Fifth Session, Aizawl, Mizoram, 1984. A paper presented by Dr. Mahadev Chakraborty, The Ziratia Problem of Tripura.

Seminar on Trans-Border Migration, University of Rajasthan, Jaipur Mar., 23-25, 1987. A Paper presented by K. Singh, India and the South Asia.

Seminar held by the Planning Forum of the Jadavpur University, 4-6 February, 1978. A Paper presented by Sri P.K. Som, Land-use Oriented Land Reform Policy.

A Discussion Paper on National Consultation on Perspective plan for Conservation, Management and Development of Land Resources of the Country, New Delhi, Oct., 21-24, 1991.

North-Eastern Council, Basic Statistics of North-Eastern Region, Shillong, 1984, 1990, 1994, 1995.

Proceedings of a Seminar Land, Land Relations and Land Reforms in the Plains and Valleys of North East India ed. by Atul Goswami, April 17-18, 1984.

Agro-Economic Research Centre, *An Assessment of Colonisation Scheme of Shifting Cultivators in Tripura*, Jorhat, 1969.

Journal of Agricultural Economics, Vol. XXV, Bombay, 1970. An article by U. Dhar, Shifting Cultivation in Hill Tracts of India.

Educational Miscellany (special Issue) Vol. XIII July, 1982-June, 1984. An article by Dr. Mahadev Chakraborti, Organisation of the Ziratia Tenants.

Hindu (The), Survey of Environment, 1992. An article by M. Gadgil and Ramchandra Guha. An Outline of Participatory System of Forest Management.

Our Environment, National Book Trust, New Delhi, 3rd ed. 1994. An article by L. Futehally, Forests.

Books

Agarwal, A.K. : *Economic Problems and Planning in North-East India*, Sterling, New Delhi, 1987.

Agarwal, A.K. : *North-Eastern Economy, Problems and Prospects*, Mittal Publications, New Delhi, 1988.

Aitchinson, C.U., *A Collection of Treatise, Engagements and Sanads*, Appendix VII, Vol. I, 1892.

Barkataki, S. (ed.), *Tribes of Assam*, National Book Trust, India, 1969.

Bhattacharjee, P.R., *Economic Transition in Tripura*, Vikas Publishing House, New Delhi, 1993.

Bhattacharjee, S., *Genesis of Tribal Extremism in Tripura*, Gyan Publishing House, Delhi, 1992.

Bhattacharjee, S.R., *Tribal Insurgency in Tripura–A Study in Exploration of Causes*, Inter-India Publication, New Delhi, 1989.

Bhattacharjya, Gayatri, *Refugee Rehabilitation and its Impact on Tripura's Economy*, Omsons Publications, Gowahati, 1988.

Bhattacharya, S.C., *Tripura on the March, Public Relation*, Office, Agartala, 1967.

Boserup, E., *The Conditions of Agricultural Growth*, Aldine Publishing Co. Chicago, 1965.

Chakraborty, M., Ganguly, A. and Das Gupta, M.(ed.), *Development of Forestry in Tripura* (Compilation of Seminar Papers of NEICSSR) Omson's 1989.

Chandra, Bipan, *Modern India*, National Council of Educational Research and Training, New Delhi, 1976.

Chatterjee, P., *The Land Question of Bengal, 1920-1947*, Vol. I, K.P. Bagchi and Co., Calcutta, 1984.

Choudhury, D.K. (ed.), *Annual Administrative Repots of The Political Agency, 1872-79, 1889-90*, Vol. I and II. Tribal Research Centre, Agartala, 1994.

Conklin, H.C., The *Study of Shifting Cultivation, Honuroo Agriculture in Phillippines*, F.A.O. Rome, 1951.

Das, J.N., *A Study of the Land System of Tripura*, Gauhati Law Research Institute, 1990.

Davis, W.M., *Geographical Essays*, New York, (Reprint) 1954.

De, N.K., *Measuring Land Potentials in Developing Countries*, University of Burdwan, 1984.

Deb Barma, Somendra Chandra, *Census Bibarani*, Agartala, 1931.

Dutta, B.C., *Thirty Years in Tripura (Dharmanagar division) in Bengali*, Agartala.

Dutta, B.C., *Thirty Years in Tripura (Udaipur Bibaran) Agartala*, 1340 T.E.

Dutta, D.C. and Banerjee, S. (ed.), *Rajgi Tripurar Sarkari Bangla*, Education Directorate, Government of Tripura, 1976.

F.A.O. Staff, *Shifting Cultivation The Unasylva*, Vol. I, Rome, 1957.

Fleet, J.F., *Topographical list of Brihat Samhita*, Calcutta, 1973.

Ganguly J.B., *Economic Problems of the Jhumias of Tripura*, Bookland Pvt. Ltd. Calcutta, 1969.

Ganguly, J.B., *The Benign Hills*, Tripura Darpan Prakashani, Agartala, 1983.

Ghosh, K., *Tribal Insurrection in Tripura*, Booklinks Pvt. Ltd. Hyderabad, 1984.

Gon Choudhury, J. (ed.), *Tripura–The Land and its People*, Leela Devi Publications, Delhi, 1980.

Goswami, Atul (ed.), *Land Reforms and Peasant Movements: A Study of North-East India*, Omsons Publications, Guwahati, 1986.

Government of Tripura, Directorate of Land Records and Settlement, Land Records, *Land Reforms and Maintenance of Records in Tripura, A Hand-book on Cadastral Survey*, 1985.

Hunter, W.W., *A Statistical Account of Bengal*, Vol. VI, London 1876. (Reprinted, 1973).

Imperial Gazetteer of India (The), Vol. XIII, Oxford University Press, London, 1908.

Joerg, H.L., *The Sub-division of North America into Natural Regions–A Preliminary Enquiry, Annuals of AAG*, Washington, D.C., Vol. IV, 1914.

Jones, C.F. and Darkenwold, *Economic Geography*, New York, 1950.

Kellog, C.E., *A Method of Rural Land Classification*, USDA Technical Bulletin, No. 169.

Khan, A.M., *The Transition in Bengal 1756-65*, Cambridge University Press, London, 1969.

Mackenzie, A., *History of the Relations of the Government with the Hill Tribes of the North-Eastern Frontier of Bengal*, Home Department Press, Calcutta, 1884 (Reprinted by Mittal Publications, Delhi, 1979 & 1981).

Mahajan, V.S. (ed.), *Studies in Indian Agriculture and Rural Development*, Deep and Deep Publication, New Delhi, 1992.

Majumdar, R.C., *History of Ancient Bengal*, Calcutta, 1971.

Mathew T. (ed.), *Tribal Economy of the North-Eastern Region*, Spectrum Publications, Gauhati, 1980.

Menon, K.D., *Tripura District Gazetteer*, Education Directorate, Agartala, 1975.

Misra, B.P., *Socio-economic Adjustment of Tribals—A Case Study of Tripura Jhumiyas*, People's Publishing House, New Delhi, 1976

Nair, Sukumaran M.K., *Tribal Economy in Transition (A Study of Meghalaya)*. Inter-India Publications, New Delhi, 1987.

National Council of Applied Economic Research, *Techno-Economic Survey of Tripura*, New Delhi, 1961.

Powell, J.W., *The Physiography of the United States*, American Book Co., New York, 1896.

Ramaiah, P., *Tribal Economy of India*, Light and Life Publishers, New Delhi, 1981.

Ray Burman B.K., *Demographic and Socio-Economic Profile of Hill Areas of North East India*, Registrar General, Govt. of India, New Delhi, 1970.

Roy Choudhury, N.R., *Tripura Through the Ages*, Bureau of Research and Publication on Tripura, Agartala.

Saigal, O., *Tripura - Its History and Culture*, Concept Publishing Company, Delhi, 1978.

Sandys, E.F., *History of Tripura, Agartala*, Tripura State Government, 1915.

Sen Tripurari, *Tripura in Transition 1923-57*, Agartala, 1970.

Sharma, R.M., *Political History of Tripura*, Calcutta, 1986.

Sinha K.C., *Rajmala* (4th Lahar) Education Directorate, Tripura, 1976.

Talukdar, S.P., *The Chakma Life and Struggle*, Gyan Publishing House, Delhi, 1988.

The New Encyclopaedia Britannica, Vol. II, University of Chicago, USA, 1982.

Thomson, W.H., *Final Report on the Survey and Settlement of Tipperah*, Calcutta, 1920.

Index

M

N

O

P